NATEF Standards Job Sheets

Electrical and Electronic Systems (A6)

Second Edition

Jack Erjavec

THOMSON
DELMAR LEARNING

Australia Canada Mexico Singapore Spain United Kingdom United States

THOMSON
DELMAR LEARNING

NATEF Standards Job Sheets

Electrical and Electronic Systems (A6)
Second Edition

Jack Erjavec

Vice President, Technology and Trades SBU:
Alar Elken

Editorial Editor:
Sandy Clark

Senior Acquisitions Editor:
David Boelio

Development Editor:
Matthew Thouin

Marketing Director:
David Garza

Channel Manager:
William Lawrensen

Marketing Coordinator:
Mark Pierro

Production Director:
Mary Ellen Black

Production Editor:
Toni Hansen

Art/Design Specialist:
Cheri Plasse

Technology Project Manager:
Kevin Smith

Editorial Assistant:
Andrea Domkowski

Printed in United States of America
1 2 3 4 5 XXXX 09 08 07 06 05

For more information contact Thomson Delmar Learning, Executive Woods, 5 Maxwell Drive, Clifton Park, NY 12065-2919.
Or find us on the World Wide Web at http://www.delmarlearning.com.

ISBN: 1-4180-2079-6

CONTENTS

PREFACE

The automotive service industry continues to change with the technological changes made by automobile and tool and equipment manufacturers. Today's automotive technician must have a thorough knowledge of automotive systems and components, good computer skills, exceptional communication skills, good reasoning, the ability to read and follow instructions, and above average mechanical aptitude and manual dexterity.

This new edition, like the last, was designed to give students a chance to develop the same skills and gain the same knowledge that today's successful technician has. This edition also reflects the changes in the guidelines established by the National Automotive Technicians Education Foundation (NATEF), as of July 2005.

The purpose of NATEF is to evaluate technician training programs against standards developed by the automotive industry and recommend qualifying programs for certification (accreditation) by ASE (National Institute for Automotive Service Excellence). Programs can earn ASE certification upon the recommendation of NATEF. NATEF's national standards reflect the skills that students must master. ASE certification through NATEF evaluation ensures that certified training programs meet or exceed industry-recognized, uniform standards of excellence.

At the expense of much time and many minds, NATEF has assembled a list of basic tasks for each of their certification areas. These tasks identify the basic skills and knowledge levels that competent technicians have. The tasks also identify what is required for a student to start a successful career as a technician.

Most of the content in this book are job sheets. These job sheets relate to the tasks specified by NATEF. The main considerations during the creation of these job sheets were student learning and program certification by NATEF. Students are guided through standard industry accepted procedures. While they are progressing, they are asked to report their findings as well as offer their thoughts on the steps they have just completed. The questions asked of the students are thought provoking and require students to apply what they know to what they observe.

The job sheets were also designed to be generic. That is, whenever possible, the tasks can be performed on any vehicle from any manufacturer. Also, completion of the sheets does not require the use of specific brands of tools and equipment; rather students use what is available. In addition, the job sheets can be used as a supplement to any good textbook.

Also included are description and basic use of the tools and equipment listed in NATEF's standards. The standards recognize that not all programs have the same needs, nor do all programs teach all of the NATEF tasks. Therefore, the basic philosophy for the tools and equipment requirement is that the training should be as thorough as possible with the tools and equipment necessary for those tasks.

Theory instruction and hands-on experience of the basic tasks provide initial training for employment in automotive service or further training in any or all of the specialty areas. Competency in the tasks indicates to employers that you are skilled in that area. You need to know the appropriate theory, safety, and support information for each required task. This should include identification and use of the required tools and testing and measurement equipment required for the tasks, the use of current reference and training materials, the proper way to write work orders and warranty reports, and the storage, handling, and use of Hazardous Materials as required by the 'Right to Know Law', and federal, state, and local governments.

Words to the Instructor: I suggest you grade these job sheets based on completion and reasoning. Make sure the students answer all questions. Then look at their reasoning to see if the task was actually completed and to get a feel for their understanding of the topic. It will be easy for students to copy others' measurements and findings, but each student should have their own base of understanding and that will be reflected in their explanations.

Words to the Student: While completing the job sheets, you have a chance to develop the skills you need to be successful. When asked for your thoughts or opinions, think about what you observed. Think about what could have caused those results or conditions. You are not being asked to give accurate explanations for everything you do or observe. You are only asked to think. Thinking leads to understanding. Good technicians are good because they have a basic understanding of what they are doing and of why they are doing it.

Jack Erjavec

ELECTRICAL AND ELECTRONIC SYSTEMS

To prepare you to learn what you should learn from completing the job sheets, some basics must be covered. This discussion begins with an overview of electrical and electronic systems. Emphasis is placed on what they do and how they work. This includes the major components and designs of electrical and electronic systems and their role in the efficient operation of electrical and electronic systems of all designs.

Preparing to do something on an automobile would not be complete if certain safety issues were not addressed. This discussion covers those things you should and should not do while working on electrical and electronic systems. Included are proper ways to deal with hazardous and toxic materials.

NATEF's task list for Electrical and Electronic Systems certification is also given with defi-ni-tions of some of the terms used to describe the tasks. This list gives you a good look at what the experts say you need to know before you can be considered competent to work on electrical and electronic systems.

Following the task list are descriptions of the various tools and types of equipment you need to be familiar with. These are the tools you will use to complete the job sheets. They are also the tools NATEF has identified as being necessary for servicing electrical and electronic systems.

Following the tool discussion is a cross-reference guide that shows which NATEF tasks are related to specific job sheets. In most cases there are single job sheets for each task. Some tasks are part of a procedure and when that occurs, one job sheet may cover two or more tasks. The remainder of the book contains the job sheets.

BASIC ELECTRICAL AND ELECTRONIC SYSTEM THEORY

The electrical and electronic systems are critical parts of any automobile. In addition to playing a vital role in starting the vehicle and providing power for lighting and other auxiliary safety systems, these systems must also provide power to the sophisticated controls found on engine, brake, suspension, emission, steering, and other automotive systems.

A basic understanding of electrical principles is important to properly diagnose any system that is monitored, controlled, or operated by electricity.

Flow of Electricity

All things are made up of atoms and the basics of electricity focus in on atoms. The following principles describe atoms, which are the building blocks of all materials.

- In the center of every atom is a nucleus.
- The nucleus contains positively charged particles called protons and particles called neutrons that have no charge.
- Negatively charged particles called electrons orbit around every nucleus.
- Every type of atom has a different number of protons and electrons, but each atom has an equal number of protons and electrons. Therefore, the total electrical charge of an atom is zero, or neutral.

The looseness or tightness of the electrons in orbit around the neutron of an atom explains the behavior of electricity. Electricity is caused by the flow of electrons from one atom to another. The release of energy as one electron leaves the orbit of one atom and jumps into the orbit of another is electricity. The key behind creating electricity is to give a reason for the electrons to move.

There is a natural attraction of electrons to protons. Electrons have a negative charge and are attracted to something with a positive charge. When an electron leaves the orbit of an atom, the atom then has a positive charge. An electron moves from one atom to another because the atom next to it appears to be more positive than the one it is orbiting around. An electrical power source provides for a more positive charge, and to allow for a continuous flow of electricity, it supplies free electrons. To have a continuous flow of electricity, three things must be present: an excess of electrons in one place, a lack of electrons in another place, and a path between the two places.

Two power or energy sources are used in an automobile's electrical system. These are based on a chemical reaction and on magnetism. A car's battery (Figure 1) is a source of chemical energy. A chemical reaction in the battery provides for an excess of electrons and a lack of electrons in another place. Batteries have two terminals: a positive and a negative. Basically, the negative terminal is the outlet for the electrons and the positive terminal is the inlet for the electrons to get to the protons. The chemical reaction in a battery causes a lack of electrons at the positive (+) terminal and excess at the negative (–) terminal. This creates an electrical imbalance, causing the electrons to flow through the path provided by a wire.

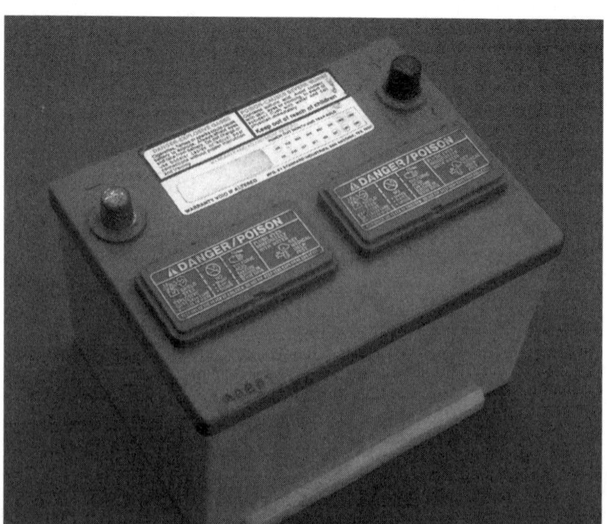

Figure 1 A typical battery.

The chemical process in the battery continues to provide electrons until the chemicals become weak. At that time, either the battery has run out of free electrons or each of the protons is matched with an electron. When this happens, there is no longer a reason for the electrons to move to the positive side of the battery. Fortunately, the vehicle's charging system frees up electrons in the battery and allows the chemical reaction in the battery to continue indefinitely.

Electricity and magnetism are interrelated. One can be used to produce the other. Moving a wire through an already existing magnetic field can produce electricity. This process of producing electricity through magnetism is called induction. In an AC generator, a magnetic field is moved through a coil of wire and voltage is induced. The amount of electricity produced depends on a number of factors, including the strength of the magnetic field, the number of wires that pass through the field, and the speed at which the wire moves through the magnetic field.

Electrical Terms

Electrical current is a term used to describe the movement or flow of electricity. The greater the number of electrons flowing past a given point in a given amount of time, the more current the circuit has. This current, like the flow of water or any other substance, can be measured. Voltage is electrical pressure. Voltage is the force developed by the attraction of the electrons to the protons. The more positive one side of the circuit is, the more voltage is present in the circuit. Voltage does not flow; rather, it is the pressure that causes current flow.

When any substance flows, it meets resistance. The resistance to electrical flow can be measured. Power is the rate at which work can be done. Forcing a current through a resistance is work. The power used in a circuit can also be determined.

Electrical Current

The unit for measuring electrical current is the ampere. There are two types of electrical flow, or current: direct current (DC) and alternating current (AC). In direct current, the electrons flow in one direction only. In alternating current, the electrons change direction at a fixed rate. An automobile uses DC current, whereas the current in homes and buildings is AC.

Resistance

In every atom, the electrons resist being moved out of their shell. The amount of resistance depends on the type of atom. In some atoms there is very little resistance to electron flow because the outer electron is loosely held. Materials made up of these types of atoms are typically referred to as conductors. In other substances, there is more resistance to flow, because the outer electrons are tightly held. These materials are called insulators.

The resistance to current flow produces heat. This heat can be measured to determine the amount of resistance. A unit of measured resistance is called an ohm.

Voltage

In electrical flow, some force is needed to move the electrons between atoms. This force is the pressure that exists between a positive and negative point within an electrical circuit. This force, also called electromotive force (EMF), is measured in units called volts. One volt is the amount of pressure (force) required to move one ampere of current through a resistance of one ohm.

Ohm's Law

Much of the understanding of the behavior of electricity is based on Ohm's Law. Ohm's Law is a statement and definition about the relationship between voltage, current, and resistance. This relationship is expressed in a mathematical formula that says there must be one volt of electrical pressure to push one ampere of electrical current through one ohm of electrical resistance. With this formula, the behavior of electricity can be predicted.

We know, from Ohm's Law, that when resistance in a circuit increases, circuit current decreases. And when circuit resistance decreases, current increases. Therefore, the resistance in that circuit determines the amount of current that flows in a circuit.

The energy used by a load is measured in volts. Amperage stays constant in a circuit, but the voltage is dropped as it powers a load. Measuring voltage drop determines the amount of energy consumed by the load.

Circuits

When electrons are able to flow along a path (wire) between two points, an electrical circuit is formed. An electrical circuit is considered complete when there is a path that connects the positive and negative terminals of the electrical power source. Somewhere in the circuit there must be a load or resistance to control the amount of current in the circuit. Most automotive electrical circuits use the chassis as the path to the negative side of the battery. Electrical components have a lead that connects them to the chassis. These are called the chassis ground connections.

The metal frame acts as the return wire in the circuit. Current passes from the battery, through the load, and into the frame. The frame is connected to the negative terminal of the battery through the battery's ground wire. This completes the circuit. An electrical component, such as a generator, is often mounted directly to the engine block, transmission case, or frame. This direct mounting effectively grounds the component without the use of a separate ground wire. In other cases, however, a separate ground wire must be run from the component to the frame or another metal part to ensure a sound return path. The increased use of plastics and other nonmetallic materials in body panels and engine parts has made electrical grounding more difficult. To assure good grounding back to the battery, some manufacturers now use a network of common grounding terminals and wires.

In a complete circuit, the flow of electricity can be controlled and applied to do useful work, such as cause a headlamp to light or turn over a starter motor. Components that use electrical power put a load on the circuit and change electrical energy into another form of energy, such as heat energy.

Most automotive circuits contain four basic parts: (1) power sources, such as a battery or generator that provides the energy needed to create electron flow; (2) conductors, such as copper wires that provide a path for current flow; (3) loads, which are devices that use electricity to perform work, such as light bulbs, electric motors, or resistors; and (4) controllers, such as switches or relays that direct the flow of electrons.

Conductors and Insulators

Controlling and routing the flow of electricity requires the use of materials known as conductors and insulators. Conductors are materials with a low resistance to the flow of current. If the number of electrons in the outer shell or ring of an atom is less than 4, the force holding them in place is weak. The voltage needed to move these electrons and create current flow is relatively small. Most metals, such as copper, silver, and aluminum are excellent conductors.

When the number of electrons in the outer ring is greater than 4, the force holding them in orbit is very strong, and very high voltages are needed to dislodge them. These materials are known as insulators. They resist the flow of current. Thermal plastics are the most common electrical insulators used today. They can resist heat, moisture, and corrosion without breaking down.

Resistors

Automotive electrical circuits contain a number of different types of electrical devices. Resistors are used to limit current flow (and thereby voltage) in circuits where full current flow and voltage are not needed. Resistors are devices specially constructed to introduce a measured amount of electrical resistance into a circuit. In addition, some other components use resistance to produce heat and even light. An electric window defroster is a specialized type of resistor that produces heat. Electric lights are resistors that get so hot they produce light.

Three different types of resistors are found in automobiles: fixed value, stepped or tapped, and variable. Fixed value resistors are designed to have only one rating, which should not change. These resistors are used to control voltage such as in an automotive ignition system. Tapped or stepped resistors are designed to have two or more fixed values, available by connecting wires to the several taps of the resistor. Heat motor resistor packs, which provide for different fan speeds, are an example of this type of resistor. Variable resistors are designed to have a range of resistances available through two or more taps and a control.

Three commonly used variable resistors are rheostats, potentiometers, and thermistors. Rheostats have two connections, one to the fixed end of a resistor and one to a sliding contact with the resistor. Turning the control moves the sliding contact away from or toward the fixed end tap, increasing or decreasing the resistance. Potentiometers have three connections, one at each end of the resistance and one connected to a sliding contact with the resistor. Turning the control moves the sliding contact away from one end of the resistance, but toward the other end. Thermistors are designed to change their resistance values in response to changes in temperature. Thermistors are used to provide compensating voltage in components or to determine temperature. As a temperature sender, the thermistor is connected to a voltmeter calibrated in degrees. As the temperature rises or falls, the resistance also changes. This changes the reading on the meter.

Circuit Protection Devices

When overloads or shorts in a circuit cause too much current to flow, the wiring in the circuit heats up, the insulation melts, and a fire can result, unless the circuit has some kind of protective device. Fuses, fuse links, maxi-fuses, and circuit breakers are designed to provide protection from high current. They may be used singly or in combination.

Fuses are rated according to the current at which they are designed to blow. The current rating for blade fuses is indicated by the color of the plastic case. In addition, it is usually marked on the top. Fuse or fusible links are used in circuits where maximum current controls are not so critical. They are often installed in the positive battery lead that powers the ignition switch and other circuits that are live with the key off. A fuse link is a short length of small-gauge wire installed in a conductor. Because the fuse link is a lighter gauge of wire than the main conductor, it melts and opens the circuit before damage can occur in the rest of the circuit.

Often a maxi-fuse is used instead of a fusible link. Maxi-fuses look and operate like two-prong, blade or spade fuses, except they are much larger and can handle more current. Some circuits are protected by circuit breakers, which, like fuses, are rated in amperes. A circuit breaker conducts current through an arm made of two types of metal bonded together (bimetal arm). If the arm starts to carry too much current, it heats up. As one metal expands faster than the other, the arm bends; that opens the contacts and the current flow is broken.

Voltage Limiter

Some instrument panel gauges are protected against heavy voltage fluctuations that could damage the gauges or give erroneous readings. A voltage limiter restricts voltage to the gauges to approximately 5 volts. The limiter contains a heating coil, a bimetal arm, and a set of contacts. When the ignition is in the on or accessory position, the heating coil heats the bimetal arm, causing it to bend and open the contacts. This action results in voltage from both the heating coil and the circuit. When the arm cools down to the point that the contacts close, the cycle is repeated. The rapid opening and closing of the contacts produces a pulsating voltage at the output terminal averaging about 5 volts.

Switches

A switch of some type usually controls electrical circuits. Switches turn the circuit on or off or they are used to direct the flow of current in a circuit. Switches can be under the control of the driver or can be self-operating through a condition of the circuit, the vehicle, or the environment.

A temperature-sensitive switch usually contains a bimetallic element heated either electrically or by some component where the switch is used as a sensor. When engine coolant is below or at normal operating temperature, the engine coolant temperature sensor is in its normally open condition. If the coolant exceeds the temperature limit, the bimetallic element bends the two contacts together and the switch is closed to the indicator or the instrument panel. Other applications for heat-sensitive switches are time delay switches and flashers.

Relays

A relay is an electric switch that allows a small amount of current to control a much larger one. When the control circuit switch is open, no current flows to the coil of the relay, so the windings are deenergized. When the switch is closed, the coil is energized, turning the soft iron core into an electromagnet and drawing the armature down. This closes the power circuit contacts, connecting power to the load circuit. When the control switch is opened, the current stops slowing in the coil, the electromagnetic field disappears, and the armature is released, which breaks the power circuit contacts.

Solenoids

Solenoids are also electromagnets with movable cores used to translate electrical current flow into mechanical movement. They can also close contacts, acting as a relay at the same time.

Capacitors (Condensers)

Capacitors are constructed from two or more sheets of electrically conducting material with a nonconducting or dielectric material placed between them and conductors connected to the two sheets. Capacitors are typically used as filters to remove spikes and noise from voltage signals.

Electromagnetism Basics

Electricity and magnetism are related. Current flowing through a wire creates a magnetic field around the wire. Moving a wire through a magnetic field creates current flow in the wire. Many automotive components, such as generators, ignition coils, starter solenoids, and magnetic pulse generators operate using principles of electromagnetism.

A magnet has two points of maximum attraction, one at each end of the magnet. These points are called poles, with one being designated the North pole and the other the South pole. When two magnets are brought together, opposite poles attract, while similar poles repel each other.

If a straight piece of conducting wire with the terminals of a voltmeter attached to both ends is moved across a magnetic field, the voltmeter registers a small voltage reading. A voltage has been induced in the wire. The conducting wire must cut across the flux lines to induce a voltage. Moving the wire parallel to the lines of flux does not induce voltage. As a result of this induction process, the wire or conductor becomes a source of electricity and has a polarity or distinct positive and negative end. However, this polarity can be switched depending on the relative direction of movement between the wire and magnetic field. That is why an AC generator produces alternating current.

Basics of Electronics

Simply put, electronics is a technology used to control electricity. Electronics has become a special technology beyond electricity. Transistors, diodes, semiconductors, integrated circuits, and solid-state devices are all considered to be part of electronics rather than just electrical devices. But keep in mind that all the basic laws of electricity apply to electronic controls.

A semiconductor is a material or device that can function as either a conductor or an insulator, depending on how its structure is arranged. Semiconductor materials have less resistance than an insulator but more resistance than a conductor. Because semiconductors have no moving parts, they seldom wear out or need adjustment. Semiconductors are also small, require little power to operate, are reliable, and generate very little heat. For all these reasons, semiconductors are being used in many applications.

Because a semiconductor can function as both a conductor and an insulator, it is very useful as a switching device. How a semiconductor functions depends upon the way current flows (or tries to flow) through it. Two common semiconductor devices are diodes and transistors.

Diodes

The diode is the simplest semiconductor device. A diode allows current to flow in one direction, but not in the opposite direction. Therefore, it can function as a switch, acting as either a conductor or insulator, depending on the direction of current flow. The most commonly used diodes are regular diodes, LEDs, zener diodes, clamping diodes, and photo diodes.

One application of diodes is in the AC generator, where they function as one-way valves for current flow. Generators produce alternating current. The AC must be changed to DC before it is sent to the vehicle's electrical circuit. Diodes in the generator are arranged so that current can leave the generator in one direction only (as direct current).

A variation of the diode is the zener diode. This device functions like a standard diode until a certain voltage is reached. When the voltage level reaches this point, the zener diode allows current to flow in the reverse direction. Zener diodes are often used in electronic voltage regulators.

Whenever the current flow through a coil of wire (such as used in a solenoid or relay) stops, a voltage surge or spike is produced. This surge results from the collapsing of the magnetic field around the coil. The movement of the field across the winding induces a very high voltage spike, which can damage electronic components. In the past, a capacitor was used as a "shock absorber" to prevent component damage from this surge. On today's vehicles, a clamping diode is commonly used to prevent this voltage spike. By installing a clamping diode in parallel to the coil, a bypass is provided for the electrons during the time the circuit is opened.

An example of the use of clamping diodes is on some air-conditioning compressor clutches. Because the clutch operates by electromagnetism, opening the clutch coil circuit produces a voltage spike. If the spike was left unchecked, it could damage the clutch coil relay contacts or the vehicle's computer.

Transistors

A transistor is an electronic device produced by joining three sections of semiconductor materials. Like the diode, it is very useful as a switching device, functioning as either a conductor or an insulator. Transistors are also used as signal amplifiers in radios, stereos, calculators, computers, and computerized engine controls.

Integrated Circuits

An integrated circuit is simply a large number of diodes, transistors, and other electronic components such as resistors and capacitors, all mounted on a single piece of semiconductor material. These circuits are very small and technology is making them smaller and more complex. The increasingly small size of integrated circuits is very important to automobiles. This means that electronics is no longer confined to simple tasks such as rectifying generator current. Transistors, diodes, and other electronic components are installed in cars to make logic decisions and issue commands to other components of the vehicle.

Electronic Circuits

A typical electronic control system is made up of sensors, actuators, and related wiring that is tied into a central processor called a microprocessor or computer. Most input sensors are designed to produce a voltage signal that varies within a given range (from high to low, including all points in between). A signal of this type is called an analog signal. Unfortunately, the computer doesn't understand analog signals. It can only read a digital binary signal, which is a signal that has only two values—on or off.

To overcome this communication problem, all analog voltage signals are converted to a digital format by an analog-to-digital converter (A/D converter). Some sensors like the Hall-effect switch produce a digital or square wave signal that can go directly to the microcomputer as input. The term *square wave* is used to describe the appearance of a digital circuit after it has been plotted on a graph. The abrupt changes in circuit condition (on and off) result in a series of horizontal and vertical lines that connect to form a square-shaped pattern.

In addition to A/D conversion, some voltage signals require amplification before they can be relayed to the computer. To perform this task, an input conditioner known as an amplifier is used to strengthen weak voltage signals.

After input has been generated, conditioned, and passed along to the microcomputer, it is ready to be processed for the purposes of performing work and displaying information. The portion of the microcomputer that receives sensor input and handles all calculations (makes decisions) is called the microprocessor. In order for the microprocessor to make the most informed decisions regarding system operation, sensor input is supplemented by the memory.

A computer's memory holds the programs and other data, such as vehicle calibrations, which the microprocessor refers to in performing calculations. To the computer, the program is a set of instructions or procedures that it must follow. Included in the program is information that tells the microprocessor when to retrieve input (based on temperature, time, etc.), how to process the input, and what to do with it once it has been processed.

The microprocessor works with memory in two ways: it can read information from memory or change information in memory by writing in or storing new information. During processing, the computer often receives more data than it can immediately handle. In these instances, some information is temporarily stored or written into memory until the microprocessor needs it.

Sensors

All sensors perform the same basic function. They detect a mechanical condition (movement or position), chemical state, or temperature condition and change it into an electrical signal that can be used by the computer to make decisions. The computer makes decisions based on information it receives from sensors and the programmed instructions it has in its memory. Each sensor used in a particular system has a specific job to do (for example, monitor throttle position, vehicle speed, manifold pressure). Together these sensors provide enough information to help the computer form a complete picture of vehicle operation.

Reference voltage (Vref) sensors provide input to the computer by modifying or controlling a constant, predetermined voltage signal. This signal, which can have a reference value from 5 to 9 volts, is generated and sent out to each sensor by a reference voltage regulator located inside the processor. Because the computer knows that a certain voltage value has been sent out, it can indirectly interpret things like motion, temperature, and component position, based on what comes back. Variable resistors are typically used as voltage-reference sensors.

Two other commonly used reference voltage sensors are switches and thermistors. Switches tell the computer when something is turned on or off and when a particular condition exists. Switches don't provide a variable signal; there is either a signal from them, or there is no signal. Thermistors are temperature sensitive and send a varying signal to the computer based on the temperature they are subject to.

Voltage generating sensors include components like the Hall-effect switch, oxygen sensor (zirconium dioxide), and knock sensor (piezoelectric), which are capable of producing their own input voltage signal. This varying voltage signal, when received by the computer, enables the computer to monitor and adjust for changes in the operation of various systems of the automobile.

Actuators

After the computer has assimilated the information and the tools used by it to process this information, it sends output signals to control devices called actuators. These actuators are solenoids, switchers, relays, or motors, which physically act or carry out a decision the computer has made.

Actuators are electromechanical devices that convert an electrical current into mechanical action. This mechanical action can then be used to open and close valves, control vacuum to other components, or open and close switches. When the computer receives an input signal indicating a change in one or more of the operating conditions, the computer determines the best strategy for handling the conditions. The computer then controls a set of actuators to achieve a desired effect or strategy goal. In order for the computer to control an actuator, it must rely on a component called an output driver.

Output drivers are located in the processor and operate by the digital commands issued by the computer. Basically, the output driver is nothing more than an electronic on/off switch that the computer uses to control the ground circuit of a specific actuator.

For actuators that cannot be controlled by a solenoid, relay, switches, or motors, the computer must turn its digitally coded instructions back into an analog format via a digital-to-analog converter.

Multiplexing

Multiplexing is an in-vehicle networking system used to transfer data between electronic modules through a serial data bus. Serial data is electronically coded information that is transmitted by one computer and received and displayed by another computer. Serial data is information that is digitally coded and transmitted in a series of data bits. The data transmission rate is referred to as the baud rate. Baud rate refers to the number of data bits that can be transmitted in a second.

With multiplexing, fewer dedicated wires are required for each function, and this reduces the size of the wiring harness. Using a serial data bus reduces the number of wires by combining the signals on a single wire through time division multiplexing. Information is sent to individual control modules that control each function, such as antilock braking, turn signals, power windows, and instrument panel.

Multiplexing also eliminates the need for redundant sensors because the data from one sensor is available to all electronic modules. Multiplexing also allows for greater vehicle content flexibility because functions can be added through software changes, rather than adding another module or modifying an existing one.

Batteries

The storage battery is the heart of a vehicle's electrical and electronic systems. It plays an important role in the operation of the starting, charging, ignition, and accessory circuits. The storage battery converts electrical current from the generator into chemical energy, and then stores that energy until it is needed. When switched into an external electrical circuit, the battery's chemical energy is converted back to electrical energy.

A vehicle's battery has three main functions. It provides voltage and serves as a source of current for starting, lighting, and ignition. It acts as a voltage stabilizer for the entire electrical system of the vehicle. And, finally, it provides current whenever the vehicle's electrical demands exceed the output of the charging system.

Battery Voltage and Capacity

The open-circuit voltage of a fully charged battery cell is roughly 2.1 volts. Therefore a fully charged battery has at least 12.6-volts. Cell size, state of charge, rate of discharge, battery condition and design, and electrolyte temperature all strongly influence the voltage of a battery during discharge. When cranking an engine over at 80°F, the voltage of an average battery may be about 11.5 to 12 volts. At 0°F, the voltage is significantly lower.

The concentration of acid in the electrolyte in the pores of the plates also affects battery voltage or discharge. As the acid chemically combines with the active materials in the plates and is used up, the voltage drops unless fresh acid from outside the plate moves in to take its place. As discharging continues, this outside acid becomes weaker, and sulfate saturates the plate material. It then becomes increasingly difficult for the chemical reaction to continue and, as a result, voltage drops to a level no longer effective in delivering sufficient current to the electrical system.

Battery capacity is the ability to deliver a given amount of current over a period of time. It depends on the number and size of the plates used in the cells and the amount of acid used in the electrolyte. Batteries are rated according to reserve capacity and cold-cranking power.

Starting System

The starting system is designed to turn or crank the engine over until it can operate under its own power. To do this, the starter motor receives electrical power from the battery. The starter motor then converts this energy into mechanical energy, which it transmits through the drive mechanism to the engine's flywheel.

A typical starting system has six basic components and two distinct electrical circuits. The components are the battery, ignition switch, battery cables, magnetic switch (either electrical relay or solenoid), starter motor, and the starter safety switch.

The starting system is designed with two connected circuits: the starter circuit and the control circuit. The starter circuit carries heavy current flow from the battery to the starter motor by way of the relay or solenoid. The control circuit controls the action of the solenoid and/or relay. The control circuit uses low current to control the high current starter circuit.

The starting circuit requires two or more heavy-gauge cables that attach directly to the battery. One of these cables makes a connection between the battery's negative terminal and a good chassis ground. The other cable connects the battery's positive terminal with the relay or solenoid. On vehicles where the magnetic switch is not mounted directly on the starter motor, the positive cable is actually two cables. One runs from the positive battery terminal to the switch and the second from the switch to the starter motor terminal.

Every starting system contains some type of magnetic switch that enables the control circuit to open and close the starter circuit. A solenoid-actuated starter is the most common starter system used. In this system, the solenoid mounts directly on top of the starter motor. The solenoid uses the electromagnetic field generated by its coil to perform two distinct jobs. It pushes the drive pinion of the starter motor into mesh with the engine flywheel. This is its mechanical function. Secondly, it acts as an electrical relay switch to energize the motor once the drive pinion is engaged. Once the contact points of the solenoid are closed, full current flows from the battery to the starter motor.

The solenoid assembly has two separate windings: a pull-in winding and a hold-in winding. The two windings have approximately the same number of turns but are wound from different size wire. Together these windings produce the electromagnetic force needed to pull the plunger into the solenoid coil. The heavier pull-in windings draw the plunger into the solenoid, while the lighter gauge windings produce enough magnetic force to hold the plunger in this position.

Both windings are energized when the ignition switch is turned to the start position. When the plunger disc makes contact with the solenoid terminals, the pull-in winding is deactivated. At the same time, the plunger contact disc completes the connection between the battery and the starting motor, directing full battery current to the field coils and starter motor armature for cranking power.

As the solenoid plunger moves, the shift fork also pivots on the pivot pin and pushes the starter drive pinion into mesh with the flywheel ring gear. When the starter motor receives current, its armature starts to turn. This motion is transferred through an overriding clutch and pinion gear to the engine flywheel and the engine is cranked.

Starter relays are connected in series with the battery cables to deliver the high current necessary through the shortest possible battery cables. A relay serves as the control for the starter circuit. Often they are used with positive engagement starters that use the magnetic field inside the starter motor to move the drive pinion into the flywheel.

Some vehicles use both a starter relay and a starter motor mounted solenoid. The relay controls current flow to the solenoid, which in turn controls current flow to the starter motor. This reduces the amount of current flowing through the ignition switch. In other words, it takes less current to activate the relay than to activate the solenoid.

Starter Motors

A starting motor is a special type of electric motor designed to operate under great electrical overloads and to produce very high horsepower. All starting motors are generally the same in design and operation. Basically the starter motor consists of a housing, field coils, an armature, a commutator and brushes, and end frames. The main difference between designs is in the drive mechanism used to engage the flywheel.

The starter housing or frame encloses the internal starter components and protects them from damage, moisture, and foreign materials. The housing supports the field coils and forms a path for the magnetism produced by the current passing through the coils.

The field coils and their pole shoes are securely attached to the inside of the iron housing. The field coils and pole shoes are designed to produce strong stationary electromagnetic fields within the starter body as current is passed through the starter. These magnetic fields are concentrated at the pole shoe.

The field coils connect in series with the armature winding through the starter brushes. This design permits all current passing through the field coil circuit to also pass through the armature windings.

The armature is the only rotating component of the starter. When the starter operates, the current passing through the armature produces a magnetic field in each of its conductors. The reaction between the armature's magnetic field and the magnetic fields produced by the field coils causes the armature to rotate. The armature has two main components: the armature windings and the commutator.

The coils connect to each other and to the commutator so that current from the field coils flows through all of the armature windings at the same time. This action generates a magnetic field around each armature winding, resulting in a repulsion force all around the conductor. This repulsion force causes the armature to turn.

The commutator assembly is made up of heavy copper segments separated from each other and the armature shaft by insulation. The commutator segments connect to the ends of the armature windings. Most starter motors have two to six brushes that ride on the commutator segments and carry the heavy current flow from the stationary field coils to the rotating armature windings via the commutator segments.

The control circuit usually consists of an ignition switch connected through normal gauge wire to the battery and the magnetic switch (solenoid or relay). When the ignition switch is turned to the start position, a small amount of current flows through the coil of the magnetic switch, closing it and allowing full current to flow directly to the starter motor. The ignition switch performs other jobs besides controlling the starting circuit. It normally has at least four separate positions: accessory, off, on (run), and start.

The starting safety switch, often called the neutral safety switch, is a normally open switch that prevents the starting system from operating when the transmission is in gear. The safety switch used with an automatic transmission can be either an electrical switch or a mechanical device. Contact points on the electrical switch are only closed when the shift selector is in park or neutral. The safety switches used with manual transmissions are usually electrical switches activated by the movement of the clutch pedal and/or position of the shift linkage.

Charging Systems

The charging system converts the mechanical energy of the engine into electrical energy. During cranking, the battery supplies all of the vehicle's electrical energy. However, once the engine is running, the charging system is responsible for producing enough energy to meet the demands of all the loads in the electrical system, while also recharging the battery.

In an AC generator, sometimes called an alternator, a spinning magnetic field rotates inside stationary conductors. As the spinning north and south poles of the magnetic field pass the conducting wires, they induce voltage that first flows in one direction and then in the opposite direction. Because automotive electrical systems operate on direct current, this alternating current must be changed or rectified into direct current by diodes.

The rotor assembly consists of a drive shaft, coil, and two pole pieces. A pulley mounted on the shaft end allows the rotor to be spun by a belt driven from the crankshaft pulley. The rotor produces the rotating magnetic field. The magnetic field is generated by passing a small amount of current through the coil windings. As current flows through the coil, the core is magnetized and the pole pieces assume the magnetic polarity of the end of the core that they touch. Thus, one pole piece has a north polarity and the other has a south polarity. The extensions of the pole pieces form the actual magnetic poles.

Current to create the magnetic field is supplied to the coil from one of two sources, the battery, or when the engine is running, the alternator itself. In either case, the current is passed through the voltage regulator before being applied to the coil. The voltage regulator varies the amount of current supplied. Increasing field current to the coil increases the strength of the magnetic field. This, in turn, increases alternator voltage output. Decreasing the field voltage to the coil has the opposite effect. Output voltage decreases.

Slip rings and brushes conduct current to the rotor. Two slip rings are mounted directly on the rotor shaft and are insulated from the shaft and each other. Each end of the field coil connects to one of the slip rings. A carbon brush located on each slip ring carries the current to and from the field coil. Current is transmitted from the field terminal of the voltage regulator through the first brush and slip ring to the field coil. Current passes through the field coil and the second slip ring and brush before returning to ground.

The stator is the stationary member of the AC generator and is made up of a number of conductors. The stator assembly in most AC generators has three separate windings, each placed in slightly different positions so their electrical pulses are staggered.

The rotor fits inside the stator. A small air gap between the two allows the magnetic field of the rotor to energize all of the stator windings at the same time.

A functioning generator must be regulated. A voltage regulator controls the amount of current produced by the generator and thus the voltage level in the charging circuit. Without a voltage regulator, the battery would be overcharged and the voltage level in the electrical systems would rise to the point where lights would burn out and fuses and fusible links would blow.

The regulation of the charging circuit is accomplished by varying the amount of field current flowing through the rotor. Output is high when the field current is high and low when the field current is low. The operation of the regulator is comparable to that of a variable resistor in series with the field coil. If the resistance the regulator offers is low, the field current is high and if the resistance is high, the field current is low. The amount of resistance in series with the field coil determines the amount of field current, the strength of the rotor's magnetic field, and thus the amount of generator output. The resistance offered by the regulator varies according to charging system demands or needs.

In order to regulate the charging system, the regulator must have system voltage as an input. This voltage is also called sensing system voltage because the regulator is sensing system voltage. The regulator determines the need for charging current according to the level of the sensing voltage. When the sensing voltage is less than the regulator setting, the regulator increases the field current in order to increase the charging current. As sensing voltage rises, a corresponding decrease in field current and system output occurs. Thus, the regulator responds to changes in system voltage by increasing or decreasing charging current.

On a growing number of late-model vehicles, a separate voltage regulator is no longer used. Instead, the voltage regulation circuitry is located in the vehicle's electronic control module.

This type of system does not control rotor field current by acting like a variable resistor. Instead, the computer switches or pulses field current on and off at a fixed frequency of about 400 cycles per second. By varying on-off times, a correct average field current is produced to provide correct alternator output.

Headlights

In a typical headlight system, power is directly supplied by the battery to the headlight portion of the switch and through a fuse in the fuse panel to the remainder of the switch. Although one of the functions of the headlight switch is to control the headlights, the switch has many other functions.

There are many types of headlight switches. The pullout design has three positions: off, park, and head. Pulling a typical switch knob out to the park or first detent (knob catches at stop) illuminates all exterior lights except the headlights. Instrument panel lights are also illuminated in this position. Pulling the head switch knob out to the head or second detent illuminates the headlights plus all of those lights illuminated in park position. A dimmer rheostat controls the brightness of dash or instrument illumination lights. When the switch knob is rotated completely counterclockwise, the instrument panel illumination is at maximum brightness level. This position may also turn on the courtesy and dome/map lights.

Push button and rotary headlight switches have three positions: off, park, and head. When the park position is selected, power is applied to all circuits except the headlights. When the headlight position is selected, all of those lights illuminated in the park position are on. With these switches, the instructor panel lights are controlled at a separate rheostat switch.

Headlights have both a high and low beam. Switching from high beam to low beam is controlled by a dimmer switch. Most vehicles have column mounted dimmer switches that serve many different purposes. These switches may control the wipers, turn signals, and the high and low beam headlight circuits. Some dimmer switches have the additional feature of being able to energize both the high and low beams even if the headlight switch is off. These circuits are usually referred to as flash to pass.

Some vehicles are equipped with an automatic headlight dimmer system. It automatically switches the headlights from high to low beam in response to light from an approaching vehicle, or light from the taillights of a vehicle being overtaken. Major components of the system are a hi-lo beam photocell, an amplifier, a hi-lo beam relay, a dimmer switch, and a hi-lo beam range control.

Vehicles may also have an automatic headlight system that provides light-sensitive automatic on-off control of the headlights. This system consists of a light-sensitive photocell sensor/amplifier assembly and a headlight control relay. Turning the regular headlight switch on overrides the automatic system.

An automatic delayed-exit system keeps the lights on for a preselected period of time after the ignition switch is turned off.

Headlight Bulbs

Headlight systems typically consist of two or four sealed-beam tungsten or halogen headlight bulbs. On a two-headlight system, each headlight has two filaments, a high beam and a low beam. On the four-headlight system, the two outer headlights are of the two-filament type, and the two inner headlights have only one high-beam filament. Some vehicles use headlights that have a separate small light bulb enclosed in a reflector, with a separate lens in front. This type offers better light control and brightness when compared to sealed-beam types.

A halogen light contains a small quartz-glass bulb. Inside the bulb is a fuel filament surrounded by halogen gas. The small, gas-filled bulb fits within a larger metal reflector and lens element. A glass balloon sealed to the metal reflector allows the halogen bulb to be removed without danger of water or dirt damaging the optics within the light.

HID (High Intensity Discharge) headlights produce light with a longer reach and wider distribution on the road. The HID system uses high voltage arcing across electrodes inside a light bulb to emit light. The light bulb is a metal halide bulb containing xenon gas, mercury, and metal halide. When high voltage (approximately 20,000 volts) is applied to the electrodes inside the bulb, the xenon gas in the bulb emits light. As the temperature in the bulb increases, the mercury in the bulb evaporates and causes arcing in the bulb. As the temperature continues to increase, the metal halide in the mercury arc separates into metal and iodine atoms. The metal atoms emit a high intensity light.

An electronic control unit, located under each headlight, generates the high voltage that is applied to the electrodes of the bulbs to begin the illumination of the bulb. The control module also controls the amperage and voltage to the bulb to immediately provide an optimal amount of light after the bulbs have been turned ON and to enable the bulbs to continue to illuminate. The control module also has a fail-safe function to protect the vehicle and system from the high-voltage if there is a problem in the headlight system.

Other Lighting

Interior lights basically all operate the same. Whether the courtesy lights are on the door, under the seats, under the instrument panel, or on the rear interior quarter panels does not change how they are controlled. Also, whether the illumination lights are just behind the instrument panel or are also used in center consoles or door arm rests does not affect their operation. The only difference is the number of lights and variances in electrical wiring.

The rear light assembly includes the taillights, turn signal/stop/hazard lights/high-mounted stoplights, rear side marker lights, backup lights, and license plate lights.

Flashers are components of both turn and hazard systems. They contain a temperature-sensitive bimetallic strip and a heating element. The bimetallic strip is connected to one side of a set of contacts. Voltage from the fuse panel is connected to the other side. When the left turn signal switch is activated, current flows through the flasher unit to the turn signal bulbs. This current causes the heating element to emit heat, which in turn causes the bimetallic strip to bend and open the circuit. The absence of current flow allows the strip to cool and again close the circuit. This intermittent on/off interruption of current flow makes all left turn signal lights flash. Operation of the right turn is the same as the operation of the left turn signals.

The stop lights are usually controlled by a stop light switch mounted on the brake pedal arm or by a switch mounted to the master cylinder, which closes when hydraulic pressure indicates in response to depressing the brake pedal. In either case, voltage is present at the stop light switch at all times. Depressing the brake pedal causes the stop light switch contacts to close. Current can then flow to the stop light filament of the rear light assembly. These stay illuminated until the brake pedal is released.

When the vehicle is placed in reverse gear, backup lights are turned on to illuminate the area behind the vehicle. Various types of backup light switches are used depending on the type of transmission used on the vehicle. In general, vehicles with a manual transmission have a separate switch. Those with an automatic transmission use a combination neutral start/backup light switch. The combination neutral start/backup light switch used with automatic transmissions is actually two switches combined in one housing. In park or neutral, current from the ignition switch is applied through the neutral start switch to the starting system. In reverse, current from the fuse panel is applied through the backup light switch to the backup lights.

Light systems normally use one wire to the light, making use of the car body or frame to provide the ground back to the battery. Since many of the manufacturers have gone to plastic socket and mounting plates (as well as plastic body parts) to reduce weight, many lights must now use two wires to provide the ground connection. Some double-filament lights use two hot wires and a third ground wire. That is, double-filament bulbs have two contacts and two wire connections to them if grounded through the base. If not grounded through the base of the bulb, a two-filament bulb has three contacts and three wires connected to it. Single-filament bulbs may be single- or double-contact types. Single-contact types are grounded through the bulb base, while double-contact, single-filament types have two wires: one live and the other a ground.

Instrument Gauges

Gauges provide the driver with a scaled indication of the condition of a system. All gauges (analog or digital) require an input from either a sender or a sensor. Sender or sensor units change electrical resistance in response to changes or movements made by an external component. Movement may be caused by pressure against a diaphragm, heat, or motion of a float as liquid fills a fuel tank.

A typical mechanical speedometer has a drive cable attached to a gear in the transmission that turns a magnet inside a cup-shaped metal piece. The cup is attached to a speedometer needle and held at zero by a fine wire spring. As the cable rotates faster with increasing speed, magnetic forces act on the cup and force it to rotate. The speedometer needle, attached to the cup, moves up the speed scale.

Electronic speedometers respond to the input of a vehicle speed sensor. This speed signal is also used by other modules in the vehicle, including speed control, ride control module, the engine control module and others.

An oil pressure gauge indicates engine oil pressure based on input from a sensor. The sensor's resistance changes with changes in pressure. Some vehicles have an oil pressure switch that causes a warning lamp to light when pressure is low.

A coolant temperature gauge relies on a variable resistor-type sensor, such as a thermistor. Typically, with low coolant temperatures, sender resistance is high and current flow to the gauge is low. Therefore the gauge reads on the cold side. As coolant temperature increases, sender resistance decreases and current flow increases. The needle moves toward the hot end of the gauge.

Fuel level gauges rely on a float assembly in the fuel tank to monitor fuel level. The fuel sender unit may be combined with the fuel pump assembly. Typically, the sending unit is a variable resistor controlled by the level of an attached float in the fuel tank. When the fuel level is low, resistance in the sender is low and movement of the gauge indicator dial is minimal (from the empty position). When the fuel level is high, the resistance in the sender is high and movement of the gauge indicator (from the empty position) is greater.

A tachometer indicates engine revolutions per minute (rpm). Electrical impulses from the ignition module or coil are passed to the tachometer. The tachometer converts these impulses to rpm that can be read. The faster the engine rotates, the greater the number of impulses from the coil.

Windshield Wiper/Washer Systems

There are several types of windshield wiper systems. The wiper systems function to keep the windshield clear of rain, snow, and dirt. Headlight wipers are available that work in unison with conventional windshield wipers. The major components of a wiper system are the control switch, wiper motor and switch, and washer fluid pump. On systems with interval wipers, an interval governor is added to the circuit. The wiper motor produces a rotational motion, and an assembly of levers connected to an offset motor drive changes the rotational motion of the motor to an oscillation motion, which is needed for the wiper blades.

When the wiper switch is moved to one of the on positions, voltage from the fuse panel is applied through the wiper switch directly to the wiper motor on systems without interval wipers. Shutting off the switch cuts power to the wiper motor. However, the wiper motor park switch maintains power to the motor until it is in the park position.

On vehicles with interval wipers, current passes through the interval governor to the wiper motor. When the wiper switch is placed in the interval position, the signal is applied to an interval timer circuit in the governor. The time period of the interval is adjusted by a potentiometer in the wiper switch.

The washer pump is operated by holding the washer switch in the activate position. On models with interval wipers this signal is also applied to the governor. If the wiper switch is in off, an interval override circuit in the governor causes the wipers to operate at low speed until the washer switch is released. Then the wipers operate for several more cycles and park or return to interval operation. The wiper arms and blades are attached directly to the two pivot points operated by the linkage and motor.

Power Door Lock Systems

Although systems for automatically locking doors vary from one vehicle to another, the overall purpose is the same—to lock all outside doors. There are, however, several variations of door arrangements used that require slight differences in components from one system to another. As a safety precaution against being locked in a car due to an electrical failure, power locks can be manually operated.

When either the driver's or passenger's control switch is activated (either locked or unlocked), power from the fuse panel is applied through the switch to the door lock actuator motor. A rod that is part of the actuator moves up or down in the door latch assembly as required to lock or unlock the door. On some models the signal from the switch is applied to a relay that, when energized, applies an activating voltage to the door lock actuator. The door lock actuator consists of a motor and a built-in circuit breaker.

Power Trunk Release

The power trunk release system is a relatively simple electrical circuit that consists of a switch and a solenoid. When the trunk release switch is pressed, voltage is applied through the switch to the solenoid. With battery voltage on one side and ground on the other, the trunk release solenoid energizes and the trunk latch releases to open the trunk lid.

Power Windows

Obviously, the primary function of any power window system is to raise and lower windows. The systems do not vary significantly from one model to another. The major components of a typical system are the master control switch, individual window control switches, and the window drive motors. In addition, on four-door models, a window safety relay and in-line circuit breaker are also included. The master control switch provides overall system control.

Four-door model master control switches usually have four segments while two-door models have two segments. Each segment controls power to a separate window motor. Each segment actually operates as a separate independent switch. A window lock switch is included on four-door model master control switches. When open, this switch limits opening and closing of all windows to the master control switch. It is included as a safety device to prevent children from opening door windows without the driver knowing.

Power Seats

Power seats allow the driver or passenger to adjust the seat to the most comfortable position. The major components of the system are the seat control and the motors. In the four-way system the whole seat moves up or down, or forward and rearward. In the six-way system, the same movements are included plus the ability to adjust the height of the seat front or the seat rear. Two motors are usually employed to make the adjustments on the four-way systems, while three are used on a six-way system.

Power Mirror System

The power mirror system allows the driver to control both the left-hand and right-hand outside rear-view mirrors from one switch. The major components in the system are the joystick control switch and a dual motor drive assembly located in each mirror assembly.

Horns

Most horn systems are controlled by relays. When the horn button, ring, or padded unit is depressed, electricity flows from the battery through a horn lead, into an electromagnetic coil in the horn relay to the ground. A small flow of electric current through the coil energizes the electromagnet, pulling a movable arm. Electrical contacts on the arm touch, closing the primary circuit and causing the horn to sound.

Cruise (Speed) Control System

Cruise or speed control systems are designed to allow the driver to maintain a constant speed without having to apply continual foot pressure on the accelerator pedal. Selected cruise speeds are easily maintained and speed can be easily changed. Several override systems also allow the vehicle to be accelerated, slowed, or stopped. Because of the constant changes and improvements in technology, each cruise control system may be considerably different from others. The cruise control switch usually has several positions, including off/on, resume, and engage buttons.

A transducer is a device that controls the speed of the vehicle. When the transducer is engaged, it senses vehicle speed and controls a vacuum source (usually the manifold). The vacuum source is used to maintain a certain position on a servo. The speed control is sensed from the lower cable and casing assembly attached to the transmission. The servo unit is connected to the throttle by a rod or linkage, a head chain, or a bowden cable. The servo unit maintains the desired car speed by receiving a controlled amount of vacuum from the transducer. The variation in vacuum changes the position of the throttle. When a vacuum is applied, the servo spring is compressed and the throttle is positioned correctly. When the vacuum is released, the servo spring is relaxed and the system is not operating. When the brake pedal is depressed, the brake release switch disengages the system. A vacuum release valve is also used to disengage the system when the brake pedal is depressed.

Cruise control can also be obtained by using electronic components rather than mechanical components. Depending upon the vehicle manufacturer, several additional components may be used. The electronic control module is used to control the servo unit. The servo unit controls the vacuum required to control the throttle. A vehicle speed sensor is used to monitor or sense vehicle speed. The signal created is sent to the electronic control module, which has several inputs that help determine how it will affect the servo. These inputs include a brake release switch (clutch release switch), a speedometer, buffer amplifier, or generator speed sensor, and a speed control on the steering wheel (signal to control the cruise control).

SAFETY

In an automotive repair shop, there is great potential for serious accidents, simply because of the nature of the business and the equipment used. When people are careless, the automotive repair industry can be one of the most dangerous occupations. But, the chances of your being injured while working on a car are close to nil if you learn to work safely and use common sense. Safety is the responsibility of everyone in the shop.

Personal Protection

Some procedures, such as grinding, result in tiny particles of metal and dust that are thrown off at very high speeds. These metal and dirt particles can easily get into your eyes, causing scratches or cuts on your eyeball. Pressurized gases and liquids escaping a ruptured hose or hose-fitting can spray a great distance. If these chemicals get into your eyes, they can cause blindness. Dirt and sharp bits of corroded metal can easily fall down into your eyes while you are working under a vehicle.

Eye protection should be worn whenever you are exposed to these risks. To be safe, you should wear safety glasses whenever you are working in the shop. Some procedures may require that you wear other eye protection in addition to safety glasses. For example, when cleaning parts with a pressurized spray, you should wear a face shield. The face shield not only gives added protection to your eyes but also protects the rest of your face.

If chemicals such as battery acid, fuel, or solvents get into your eyes, flush them continuously with clean water. Have someone call a doctor and get medical help immediately.

Your clothing should be well fitted and comfortable but made of strong material. Loose, baggy clothing can easily get caught in moving parts and machinery. Some technicians prefer to wear coveralls or shop coats to protect their personal clothing. Your work clothing should offer you some protection but should not restrict your movement.

Long hair and loose, hanging jewelry can create the same type of hazard as loose-fitting clothing. They can get caught in moving engine parts and machinery. If you have long hair, tie it back or tuck it under a cap.

Never wear rings, watches, bracelets, and neck chains. These can easily get caught in moving parts and cause serious injury.

Always wear shoes or boots of leather or similar material with non-slip soles. Steel-tipped safety shoes can give added protection to your feet. Jogging or basketball shoes, street shoes, and sandals are inappropriate in the shop.

Good hand protection is often overlooked. A scrape, cut, or burn can limit your effectiveness at work for many days. A well-fitted pair of heavy work gloves should be worn during operations such as grinding and welding or when handling hot components. Always wear approved rubber gloves when handling strong and dangerous caustic chemicals.

Many technicians wear thin, surgical-type latex gloves whenever they are working on vehicles. These offer little protection against cuts but do offer protection against disease and grease buildup under and around your fingernails. These gloves are comfortable and are quite inexpensive.

Accidents can be prevented simply by the way you act. The following are some guidelines to follow while working in a shop. This list does not include everything you should or shouldn't do; it merely presents some things to think about.

- Never smoke while working on a vehicle or while working with any machine in the shop.
- Playing around is not fun when it sends someone to the hospital.
- To prevent serious burns, keep your skin away from hot metal parts such as the radiator, exhaust manifold, tailpipe, catalytic converter, and muffler.
- Always disconnect electric engine cooling fans when working around the radiator. Many of these will turn on without warning and can easily chop off a finger or hand. Make sure you reconnect the fan after you have completed your repairs.
- When working with a hydraulic press, make sure the pressure is applied in a safe manner. It is generally wise to stand to the side when operating the press.
- Properly store all parts and tools by putting them away in a place where people will not trip over them. This practice not only cuts down on injuries, it also reduces time wasted looking for a misplaced part or tool.

Work Area Safety

Your entire work area should be kept clean and safe. Any oil, coolant, or grease on the floor can make it slippery. To clean up oil, use commercial oil absorbent. Keep all water off the floor. Water makes smooth floors slippery and is dangerous as a conductor of electricity. Aisles and walkways should be kept clean and wide enough to allow easy movement. Make sure the work areas around machines are large enough to allow the machinery to be operated safely.

Gasoline is a highly flammable volatile liquid. Something that is flammable catches fire and burns easily. A volatile liquid is one that vaporizes very quickly. Flammable volatile liquids are potential firebombs. Always keep gasoline or diesel fuel in an approved safety can and never use gasoline to clean your hands or tools.

Handle all solvents (or any liquids) with care to avoid spillage. Keep all solvent containers closed, except when pouring. Proper ventilation is very important in areas where volatile solvents and chemicals are used. Solvent and other combustible materials must be stored in approved and designated storage cabinets or rooms with adequate ventilation. Never light matches or smoke near flammable solvents and chemicals, including battery acids.

Oily rags should also be stored in an approved metal container. When these oily, greasy, or paint-soaked rags are left lying about or are not stored properly, they can cause spontaneous combustion. Spontaneous combustion results in a fire that starts by itself, without a match.

Disconnecting the vehicle's battery before working on the electrical system, or before welding, can prevent fires caused by a vehicle's electrical system. To disconnect the battery, remove the negative or ground cable from the battery and position it away from the battery.

Know where all of the shop's fire extinguishers are located. Fire extinguishers are clearly labeled as to what type they are and what types of fire they should be used on. Make sure you use the correct type of extinguisher for the type of fire you are dealing with. A multipurpose dry chemical fire extinguisher will put out ordinary combustibles, flammable liquids, and electrical fires. Never put water on a gasoline fire because it will just cause the fire to spread. The proper fire extinguisher will smother the flames.

During a fire, never open doors or windows unless it is absolutely necessary; the extra draft will only make the fire worse. Make sure the fire department is contacted before or during your attempt to extinguish a fire.

Battery Safety

The potential dangers caused by the sulfuric acid in the electrolyte and the explosive gases generated during battery charging require that battery service and troubleshooting be conducted under absolutely safe working conditions. Always wear safety glasses or goggles when working with batteries no matter how small the job.

Sulfuric acid can also cause severe skin burns. If electrolyte contacts your skin or eyes, flush the area with water for several minutes. When eye contact occurs, force your eyelid open. Always have a bottle of neutralizing eyewash on hand and flush the affected areas with it. Do not rub your eyes or skin. Seek prompt medical attention if electrolyte contacts your skin or eyes. Call a doctor immediately.

When a battery is charging or discharging, it gives off quantities of highly explosive hydrogen gas. Some hydrogen gas is present in the battery at all times. Any flame or spark can ignite this gas, causing the battery to explode violently, propelling the vent caps at a high velocity and spraying acid over a wide area. To prevent this dangerous situation take these precautions:

- Never smoke near the top of a battery and never use a lighter or match as a flashlight.
- Remove wristwatches and rings before servicing any part of the electrical system. This helps to prevent the possibility of electrical arcing and burns.
- Even sealed, maintenance-free batteries have vents and can produce dangerous quantities of hydrogen if severely overcharged.
- When removing a battery from a vehicle, always disconnect the battery ground cable first. When installing a battery, connect the ground cable last.
- Always disconnect the battery's ground cable when working on the electrical system or engine. This prevents sparks from short circuits and prevents accidental starting of the engine.
- Always operate charging equipment in well-ventilated areas. A battery that has been overworked should be allowed to cool down; let air circulate around it before attempting to jump start the vehicle. Most batteries have flame arresters in the caps to help prevent explosions, so make sure that the caps are tightly in place.
- Never connect or disconnect charger leads when the charger is turned on. This generates a dangerous spark.
- Never lay metal tools or other objects on the battery, because a short circuit across the terminals can result.
- Always disconnect the battery ground cable before fast-charging the battery on the vehicle. Improper connection of charger cables to the battery can reverse the current flow and damage the AC generator.
- Never attempt to use a fast charger as a boost to start the engine.
- As a battery gets closer to being fully discharged, the acidity of the electrolyte is reduced, and the electrolyte starts to behave more like pure water. A dead battery may freeze at temperatures near 0°F. Never try to charge a battery that has ice

in the cells. Passing current through a frozen battery can cause it to rupture or explode. If ice or slush is visible or the electrolyte level cannot be seen, allow the battery to thaw at room temperature before servicing. Do not take chances with sealed batteries. If there is any doubt, allow them to warm to room temperature before servicing.

- As batteries get old, especially in warm climates and especially if they have lead-calcium cells, the grids start to grow. The chemistry is rather involved, but the point is that plates can grow to the point where they touch, producing a shorted cell.
- Always use a battery carrier or lifting strap to make moving and handling batteries easier and safer.
- Acid from the battery damages a vehicle's paint and metal surfaces and harms shop equipment. Neutralize any electrolyte spills during servicing.

Air Bag Safety

When service is performed on any air bag system component, always disconnect the negative battery cable, isolate the cable end, and wait for the amount of time specified by the vehicle manufacturer before proceeding with the necessary diagnosis or service. The average waiting period is two minutes, but some vehicle manufacturers specify up to ten minutes. Failure to observe this precaution may cause accidental air bag deployment and personal injury.

Replacement air bag system parts must have the same part number as the original part. Replacement parts of lesser or questionable quality must not be used. Improper or inferior components may result in inappropriate air bag deployment and injury to the vehicle occupants.

Do not strike or jar a sensor or an air bag system diagnostic monitor (ASDM). This may cause air bag deployment or make the sensor inoperative. Accidental air bag deployment may cause personal injury, and an inoperative sensor may result in air bag deployment failure, causing personal injury to vehicle occupants.

All sensors and mounting brackets must be properly torqued to ensure correct sensor operation before an air bag system is powered up. If sensor fasteners do not have the proper torque, improper air bag deployment may result in injury to vehicle occupants.

When working on the electrical system on an air-bag-equipped vehicle, use only the vehicle manufacturer's recommended tools and service procedures. The use of improper tools or service procedures may cause accidental air bag deployment and personal injury. For example, do not use 12V or self-powered test lights when servicing the electrical system on an air-bag-equipped vehicle.

Tool and Equipment Safety

Careless use of simple hand tools such as wrenches, screwdrivers, and hammers causes many shop accidents that could be prevented. Keep all hand tools grease-free and in good condition. Tools that slip can cause cuts and bruises. If a tool slips and falls into a moving part, it can fly out and cause serious injury.

Use the proper tool for the job. Make sure the tool is of professional quality. Using poorly made tools or the wrong tools can damage parts or the tool itself, or could cause injury. Never use broken or damaged tools.

Safety around power tools is very important. Serious injury can result from carelessness. Always wear safety glasses when using power tools. If the tool is electrically powered, make sure it is properly grounded. Before using it, check the wiring for cracks in the insulation, as well as for bare wires. Also, when using electrical power tools, never stand on a wet or damp floor. Never leave a running power tool unattended.

When using compressed air, safety glasses and/or a face shield should be worn. Particles of dirt and pieces of metal, blown by the high-pressure air, can penetrate your skin or get into your eyes.

Always be careful when raising a vehicle on a lift or a hoist. Adapters and hoist plates must be positioned correctly to prevent damage to the underbody of the vehicle. There are specific lift points that allow the weight of the vehicle to be evenly supported by the adapters or hoist plates. The correct lift points can be found in the vehicle's service manual. Before operating any lift or hoist, carefully read the operating manual and follow the operating instructions.

Once you feel the lift supports are properly positioned under the vehicle, raise the lift until the supports contact the vehicle. Then, check the supports to make sure they are in full contact with the vehicle. Shake the vehicle to make sure it is securely balanced on the lift, and then raise the lift to the desired working height. Before working under a car, make sure the lift's locking devices are engaged.

A vehicle can be raised off the ground by a hydraulic jack. The jack's lifting pad must be positioned under an area of the vehicle's frame or at one of the manufacturer's recommended lift points. Never place the pad under the floor pan or under steering and suspension components because they are easily damaged by the weight of the vehicle. Always position the jack so the wheels of the vehicle can roll as the vehicle is being raised.

Safety stands, also called jack stands, should be placed under a sturdy chassis member, such as the frame or axle housing, to support the vehicle after it has been raised by a jack. Once the safety stands are in position, the hydraulic pressure in the jack should be slowly released until the weight of the vehicle is on the stands. Never move under a vehicle when it is supported only by a hydraulic jack. Rest the vehicle on the safety stands before moving under the vehicle.

Parts cleaning is a necessary step in most repair procedures. Always wear the appropriate protection when using chemical, abrasive, and thermal cleaners.

Vehicle Operation

When the customer brings a vehicle in for service, certain driving rules should be followed to ensure your safety and the safety of those working around you. For example, before moving a car into the shop, buckle your safety belt. Make sure no one is near, the way is clear, and there are no tools or parts under the car before you start the engine. Check the brakes before putting the vehicle in gear. Then, drive slowly and carefully in and around the shop.

If the engine must be running while you are working on the car, block the wheels to prevent the car from moving. Place the transmission into park for automatic transmissions or into neutral for manual transmissions. Set the parking (emergency) brake. Never stand directly in front of or behind a running vehicle.

Run the engine only in a well-ventilated area to avoid the danger of poisonous carbon monoxide (CO) in the engine exhaust. CO is an odorless but deadly gas. Most shops have an exhaust ventilation system and you should always use it. Connect the hose from the vehicle's tailpipe to the intake for the vent system. Make sure the vent system is turned on before running the engine. If the work area does not have an exhaust venting system, use a hose to direct the exhaust out of the building.

HAZARDOUS MATERIALS AND WASTES

A typical shop contains many potential health hazards for those working in it. These hazards can cause injury, sickness, health impairments, discomfort, and even death. Here is a short list of the different classes of hazards:

- Chemical hazards are caused by high concentrations of vapors, gases, or solids in the form of dust.
- Hazardous wastes are those substances that result from performing a service.
- Physical hazards include excessive noise, vibration, pressures, and temperatures.
- Ergonomic hazards are conditions that impede normal and/or proper body position and motion.

There are many government agencies charged with ensuring safe work environments for all workers. These include the Occupational Safety and Health Administration (OSHA), Mine Safety and Health Administration (MSHA), and National Institute for Occupational Safety and Health (NIOSH). These, as well as state and local governments, have instituted regulations that must be understood and followed. Everyone in a shop has the responsibility for adhering to these regulations.

An important part of a safe work environment is the employees' knowledge of potential hazards. Right-to-know laws concerning all chemicals protect every employee in the shop. The general intent of right-to-know laws is to ensure that employers provide their employees with a safe working place as far as hazardous materials are concerned.

All employees must be trained about their rights under the legislation, the nature of the hazardous chemicals in their workplace, and the contents of the labels on the chemicals. All of the information about each chemical must be posted on material safety data sheets (MSDS) and must be accessible. The manufacturer of the chemical must give these sheets to its customers, if they are requested to do so. They detail the chemical composition and precautionary information for all products that can present a health or safety hazard.

Employees must become familiar with the general uses, protective equipment, accident or spill procedures, and any other information regarding the safe handling of the hazardous material. This training must be given to employees annually and provided to new employees as part of their job orientation.

A hazardous material must be properly labeled, indicating what health, fire, or reactivity hazard it poses and what protective equipment is necessary when handling each chemical. The manufacturer of the hazardous materials must provide all warnings and precautionary information, which must be read and understood by the user before use. A list of all hazardous materials used in the shop must be posted for the employees to see.

Shops must maintain documentation on the hazardous chemicals in the workplace, proof of training programs, records of accidents or spill incidents, satisfaction of employee requests for specific chemical information via the MSDS, and a general right-to-know compliance procedure manual utilized within the shop.

When handling any hazardous materials or hazardous waste, make sure you follow the required procedures for handling such material. Also wear the proper safety equipment listed on the MSDS. This includes the use of approved respirator equipment.

Some of the common hazardous materials that automotive technicians use are: cleaning chemicals, fuels (gasoline and diesel), paints and thinners, battery electrolyte (acid), used engine oil, refrigerants, and engine coolant (anti-freeze).

Many repair and service procedures generate what are known as hazardous wastes. Dirty solvents and cleaners are good examples of hazardous wastes. Something is classified as a hazardous waste if it is on the EPA list of known harmful materials or has one or more of the following characteristics.

- *Ignitability.* If it is a liquid with a flash point below 140°F or a solid that can spontaneously ignite.
- *Corrosivity.* If it dissolves metals and other materials or burns the skin.
- *Reactivity.* Any material that reacts violently with water or other materials or releases cyanide gas, hydrogen sulfide gas, or similar gases when exposed to low pH acid solutions. This also includes material that generates toxic mists, fumes, vapors, and flammable gases.
- *Toxicity.* Materials that leach one or more of eight heavy metals in concentrations greater than 100 times primary drinking water standard concentrations.

Complete EPA lists of hazardous wastes can be found in the Code of Federal Regulations. It should be noted that no material is considered hazardous waste until the shop is finished using it and is ready to dispose of it.

The following list covers the recommended procedure for dealing with some of the common hazardous wastes. Always follow these and any other mandated procedures.

Oil Recycle oil. Set up equipment, such as a drip table or screen table with a used oil collection bucket, to collect oils dripping off parts. Place drip pans underneath vehicles that are leaking fluids onto the storage area. Do not mix other wastes with used oil, except as allowed by your recycler. Used oil generated by a shop (and/or oil received from household "do-it-yourself" generators) may be burned on site in a commercial space heater. Also, used oil may be burned for energy recovery. Contact state and local authorities to determine requirements and to obtain necessary permits.

Oil filters Drain for at least 24 hours, crush, and recycle used oil filters.

Batteries Recycle batteries by sending them to a reclaimer or back to the distributor. Keeping shipping receipts can demonstrate that you have done the recycling. Store batteries in a watertight, acid-resistant container. Inspect batteries for cracks and leaks when they come in. Treat a dropped battery as if it were cracked. Acid residue is hazardous because it is corrosive and may contain lead and other toxic substances. Neutralize spilled acid, by using baking soda or lime, and dispose of it as a hazardous material.

Metal residue from machining Collect metal filings when machining metal parts. Keep them separate and recycle if possible. Prevent metal filings from falling into a storm sewer drain.

Refrigerants Recover and/or recycle refrigerants during the servicing and disposal of motor vehicle air conditioners and refrigeration equipment. It is not allowable to knowingly vent refrigerants to the atmosphere. Recovering and/or recycling during servicing must be performed by an EPA-certified technician using certified equipment and following specified procedures.

Solvents Replace hazardous chemicals with less toxic alternatives that perform equally. For example, substitute water-based cleaning solvents for petroleum-based solvent degreasers. To reduce the amount of solvent used when cleaning parts, use a two-stage process: dirty solvent followed by fresh solvent. Hire a hazardous waste management service to clean and recycle solvents. (Some spent solvents must be disposed of as hazardous waste, unless recycled properly). Store solvents in closed containers to prevent evaporation. Evaporation of solvents contributes to ozone depletion and smog formation. In addition, the residue from evaporation must be treated as a hazardous waste. Properly label spent solvents and store on drip pans or in diked areas and only with compatible materials.

Containers Cap, label, cover, and properly store aboveground outdoor liquid containers and small tanks within a diked area and on a paved impermeable surface to prevent spills from running into surface or ground water.

Other solids Store materials such as scrap metal, old machine parts, and worn tires under a roof or tarpaulin to protect them from the elements and to prevent the possibility of creating contaminated runoff. Consider recycling tires by retreading them.

Liquid recycling Collect and recycle coolants from radiators. Store transmission fluids, brake fluids, and solvents containing chlorinated hydrocarbons separately, and recycle or dispose of them properly.

Shop towels or rags Keep waste towels in a closed container marked "contaminated shop towels only." To reduce costs and liabilities associated with disposal of used towels (which can be classified as hazardous wastes), investigate using a laundry service that is able to treat the wastewater generated from cleaning the towels.

Waste storage Always keep hazardous waste separate, properly labeled, and sealed in the recommended containers. The storage area should be covered and may need to be fenced and locked if vandalism could be a problem. Select a licensed hazardous waste hauler after seeking recommendations and reviewing the firm's permits and authorizations.

NATEF TASK LIST FOR ELECTRICAL AND ELECTRONIC SYSTEMS

A. General Electrical System Diagnosis

A.1. Complete work order to include customer information, vehicle identifying information, customer concern, related service history, cause, and correction. — Priority Rating 1

A.2. Identify and interpret electrical/electronic system concern; determine necessary action. — Priority Rating 1

A.3. Research applicable vehicle and service information, such as electrical/electronic system operation, vehicle service history, service precautions, and technical service bulletins. — Priority Rating 1

A.4. Locate and interpret vehicle and major component identification numbers (VIN, vehicle certification labels, and calibration decals). — Priority Rating 1

A.5.	Diagnose electrical/electronic integrity for series, parallel, and series-parallel circuits using principles of electricity (Ohm's Law).	Priority Rating 1
A.6.	Use wiring diagrams during diagnosis of electrical circuit problems.	Priority Rating 1
A.7.	Demonstrate the proper use of a digital multimeter (DMM) during diagnosis of electrical circuit problems.	Priority Rating 1
A.8.	Check electrical circuits with a test light; determine necessary action.	Priority Rating 2
A.9.	Measure source voltage and perform voltage drop tests in electrical/electronic circuits using a voltmeter; determine necessary action.	Priority Rating 1
A.10.	Measure current flow in electrical/electronic circuits and components using an ammeter; determine necessary action.	Priority Rating 1
A.11.	Check continuity and resistances in electrical/electronic circuits and components with an ohmmeter; determine necessary action.	Priority Rating 1
A.12.	Check electrical circuits using fused jumper wires; determine necessary action.	Priority Rating 2
A.13.	Locate shorts, grounds, opens, and resistance problems in electrical/electronic circuits; determine necessary action.	Priority Rating 1
A.14.	Measure and diagnose the causes of abnormal key-off battery drain (parasitic draw); determine necessary action.	Priority Rating 1
A.15.	Inspect and test fusible links, circuit breakers, and fuses; determine necessary action.	Priority Rating 1
A.16.	Inspect and test switches, connectors, relays, solenoid solid state devices, and wires of electrical/electronic circuits; perform necessary action.	Priority Rating 1
A.17.	Remove and replace terminal end from connector.	Priority Rating 1
A.18.	Repair wiring harnesses and connectors connectors and terminal ends.	Priority Rating 1
A.19.	Repair wiring harnesses (including LAN/CAN/BUS systems).	Priority Rating 1
A.20.	Perform solder repair of electrical wiring.	Priority Rating 1
A.21.	Identify location of hybrid vehicle high voltage circuit disconnect (service plug) location and safety procedures.	Priority Rating 3

B. Battery Diagnosis and Service

B.1.	Perform battery state-of-charge test (conductance); determine necessary action.	Priority Rating 1
B.2.	Perform battery capacity test; confirm proper battery capacity for vehicle application; determine necessary action.	Priority Rating 1
B.3.	Maintain or restore electronic memory functions.	Priority Rating 2
B.4.	Inspect, clean, fill, and replace battery.	Priority Rating 1
B.5.	Perform slow/fast battery charge.	Priority Rating 2
B.6.	Inspect and clean battery cables, connectors, clamps, and hold-down; repair or replace as needed.	Priority Rating 1
B.7.	Start a vehicle using jumper cables and a battery or auxiliary power supply.	Priority Rating 1
B.8.	Identify high voltage circuits of electric or hybrid electric vehicle and related safety precautions.	Priority Rating 3
B.9.	Identify electronic modules, security systems and/or radios that require reinitialization or code entry following battery disconnect.	Priority Rating 2
B.10.	Identify hybrid vehicle auxiliary (12v) battery service, repair and test procedures.	Priority Rating 3

C. Starting System Diagnosis and Repair

C.1.	Perform starter current draw tests; determine necessary action.	Priority Rating 1
C.2.	Perform starter circuit voltage drop tests; determine necessary action.	Priority Rating 1
C.3.	Inspect and test starter relays and solenoids; determine necessary action.	Priority Rating 2

C.4. Remove and install starter in a vehicle. Priority Rating 2
C.5. Inspect and test switches, connectors, and wires of starter control circuits; perform necessary action. Priority Rating 2
C.6. Differentiate between electrical and engine mechanical problems that cause a slow-crank or no-crank condition. Priority Rating 1

D. Charging System Diagnosis and Repair

D.1. Perform charging system output test; determine necessary action. Priority Rating 1
D.2. Diagnose charging system for the cause of undercharge, no-charge, and overcharge conditions. Priority Rating 1
D.3. Inspect, adjust, or replace generator (alternator) drive belts, pulleys, and tensioners; check pulley and belt alignment. Priority Rating 1
D.4. Remove, inspect, and install generator (alternator). Priority Rating 1
D.5. Perform charging circuit voltage drop tests; determine necessary action. Priority Rating 1

E. Lighting Systems Diagnosis and Repair

E.1. Diagnose the cause of brighter than normal, intermittent, dim, or no light operation; determine necessary action. Priority Rating 1
E.2. Inspect, replace, and aim headlights and bulbs. Priority Rating 2
E.3. Inspect and diagnose incorrect turn signal or hazard light operation; perform necessary action. Priority Rating 2
E.4. Identify system voltage and safety precautions associated with high intensity discharge headlights. Priority Rating 3

F. Gauges, Warning Devices, and Driver Information Systems Diagnosis and Repair

F.1. Inspect and test gauges and gauge sending units for cause of intermittent, high, low, or no gauge readings; determine necessary action. Priority Rating 1
F.2. Inspect and test connectors, wires, and printed circuit boards of gauge circuits; determine necessary action. Priority Rating 3
F.3. Diagnose the cause of incorrect operation of warning devices and other driver information systems; determine necessary action. Priority Rating 1
F.4. Inspect and test sensors, connectors, and wires of electronic (digital) instrument circuits; determine necessary action. Priority Rating 3

G. Horn and Wiper/Washer Diagnosis and Repair

G.1. Diagnose incorrect horn operation; perform necessary action. Priority Rating 3
G.2. Diagnose incorrect wiper operation; diagnose wiper speed control and park problems; perform necessary action. Priority Rating 3
G.3. Diagnose incorrect windshield washer operation; perform necessary action. Priority Rating 3

H. Accessories Diagnosis and Repair

H.1. Diagnose incorrect operation of motor-driven accessory circuits; determine necessary action. Priority Rating 2
H.2. Diagnose incorrect heated glass, mirror, or seat operation; determine necessary action. Priority Rating 2
H.3. Diagnose incorrect electric lock operation; determine necessary action. Priority Rating 2
H.4. Diagnose incorrect operation of cruise control systems; repair as needed. Priority Rating 3
H.5. Diagnose supplemental restraint system (SRS) concerns; determine necessary action. (*Note*: Follow manufacturer's safety procedures to prevent accidental deployment.) Priority Rating 1
H.6. Disarm and enable the air bag system for vehicle service. Priority Rating 1
H.7. Diagnose radio static and weak, intermittent, or no radio reception; determine necessary action. Priority Rating 3
H.8. Remove and reinstall door panel. Priority Rating 1

H.9.	Diagnose body electronic system circuits using a scan tool; determine necessary action.	Priority Rating 2
H.10.	Check for module communication (LAN/CAN/BUS) errors using a scan tool.	Priority Rating 3
H.11.	Diagnose the cause of false, intermittent, or no operation of anti-theft system.	Priority Rating 2

DEFINITIONS OF TERMS USED IN THE TASK LIST

To clarify the intent of these tasks, NATEF has defined some of the terms used in the task listings. To get a good understanding of what the task includes, refer to this glossary while reading the task list.

adjust	To bring components to specified operational settings.
analyze	To examine the relationship of components of an operation.
assemble (reassemble)	To fit together the components of a device.
CAN/LAN/BUS	Refers to a high-speed communications system designed to link computers and other devices together. In-vehicle networking involving time division is also known as multiplexing. It is a method for transferring data among electronic modules via a serial data bus. Serial data is electronically coded information, which is transmitted by one computer and received and displayed by another computer. The term serial data implies that the information is digitally coded and transmitted in a series of data words.
check	To verify condition by performing an operational or comparative examination.
clean	To rid components of extraneous matter for the purpose of reconditioning, repairing, measuring and reassembling.
conductance test	A battery test used to evaluate the condition of the battery. The conductance test equipment applies a known AC voltage with a given frequency across the battery and measures the resulting current flow. The tester uses this information to evaluate the relative health of the battery as compared to the battery rating (e.g. its CCA).
determine	To establish the procedure to be used to effect the necessary repair.
determine necessary action	Indicates that the diagnostic routine(s) is the primary emphasis of a task. The student is required to perform the diagnostic steps and communicate the diagnostic outcomes and corrective actions required addressing the concern or problem. The training program determines the communication method (worksheet, test, verbal communication, or other means deemed appropriate) and whether the corrective procedures for these tasks are actually performed.
diagnose	To locate the root cause or nature of a problem by using the specified procedure.
disassemble	To separate a component's parts as a preparation for cleaning, inspection or service.
fill (refill)	To bring fluid level to a specified point or volume.
find	To locate a particular problem, e.g., shorts, grounds or opens in an electrical circuit.
high voltage	Voltages of 50 volts or higher.
identify	To establish the identity of a vehicle or component prior to service; to determine the nature or degree of a problem.
inspect	(see *check*)
install (reinstall)	To place a component in its proper position in a system.

jump-start	To use an auxiliary power supply, i.e., battery, battery charger, etc. to assist a battery to crank an engine.
listen	To use audible clues in the diagnostic process; to hear the customer's description of a problem.
locate	Determine or establish a specific spot or area.
measure	To compare existing dimensions to specified dimensions by the use of calibrated instruments and gauges.
mount	To attach or place tool or component in proper position.
parasitic draw	Electrical loads that are still present when the circuit is turned OFF.
perform	To accomplish a procedure in accordance with established methods and standards.
perform necessary action	Indicates that the student is to perform the diagnostic routine(s) and perform the corrective action item. Where various scenarios (conditions or situations) are presented in a single task, at least one of the scenarios must be accomplished.
priority ratings	Indicates the minimum percentage of tasks, by area, a program must include in its curriculum in order to be certified in that area.
reassemble	(see *assemble*)
refill	(see *fill*)
remove	To disconnect and separate a component from a system.
repair	To restore a malfunctioning component or system to operating condition.
replace	To exchange an unserviceable component with a new or rebuilt component; to reinstall a component.
reset	(see *set*)
set	To adjust a variable component to a given, usually initial, specification.
test	To verify condition through the use of meters, gauges, or instruments.
verify	To establish that a problem exists after hearing the customer's complaint and performing a preliminary diagnosis.
voltage drop	A reduction in voltage (electrical pressure) caused by the resistance in a component or circuit.

ELECTRICAL AND ELECTRONIC SYSTEMS TOOLS AND EQUIPMENT

Many different tools and many kinds of testing and measuring equipment are used to service electrical and electronic systems. NATEF has identified many of these and has said an Electrical/Electronic technician must know what they are and how and when to use them. The tools and equipment listed by NATEF are covered in the following discussion. Also included are the tools and equipment you will use while completing the job sheets. Although you need to be more than familiar with and will be using common hand tools, they are not part of this discussion. You should already know what they are and how to use and care for them.

Circuit Tester

Circuit testers are used to identify short and open circuits in any electrical circuit. Low-voltage testers are used to troubleshoot 6- to 12-volt circuits. A circuit tester, commonly called a test light, looks like a stubby ice pick. Its handle is transparent and contains a light bulb. A probe extends from one end of the handle and a ground clip and wire from the other end. When the ground clip is attached to a good ground and the probe touched to a live connector, the bulb in the handle will light up. If the bulb does not light, voltage is not available at the connector.

WARNING: *Do not use a conventional 12-V test light to diagnose components and wires in elec-*

tronic systems. The current draw of these test lights may damage computers and system components. High-impedance test lights are available for diagnosing electronic systems.

A self-powered test light is called a continuity tester. It is used on non-powered circuits. It looks like a regular test light, except that it has a small internal battery. When the ground clip is attached to the negative side of a component and the probe touched to the positive side, the lamp will light if there is continuity in the circuit. If an open circuit exists, the lamp will not light. Do not use any type of test light or circuit tester to diagnose automotive air bag systems.

Voltmeter

A voltmeter has two leads: a red positive lead and a black negative lead. The red lead should be connected to the positive side of the circuit or component. The black should be connected to ground or to the negative side of the component. Voltmeters should be connected across the circuit being tested.

The voltmeter measures the voltage available at any point in an electrical system. A voltmeter can also be used to test voltage drop across an electrical circuit, component, switch, or connector. A voltmeter can also be used to check for proper circuit grounding.

Ohmmeter

An ohmmeter measures the resistance to current flow in a circuit. In contrast to the voltmeter, which uses the voltage available in the circuit, the ohmmeter is battery powered. The circuit being tested must be open. If the power is on in the circuit, the ohmmeter will be damaged.

The two leads of the ohmmeter are placed across or in parallel with the circuit or component being tested. The red lead is placed on the positive side of the circuit and the black lead is placed on the negative side of the circuit. The meter sends current through the component and determines the amount of resistance based on the voltage dropped across the load. The scale of an ohmmeter reads from zero to infinity. A zero reading means there is no resistance in the circuit and may indicate a short in a component that should show a specific resistance. An infinite reading indicates a number higher than the meter can measure. This usually is an indication of an open circuit.

Ohmmeters are also used to trace and check wires or cables. Assume that one wire of a four-wire cable is to be found. Connect one probe of the ohmmeter to the known wire at one end of the cable and touch the other probe to each wire at the other end of the cable. Any evidence of resistance, such as meter needle deflection, indicates the correct wire. Using this same method, you can check a suspected defective wire. If resistance is shown on the meter, the wire is sound. If no resistance is measured, the wire is defective (open). If the wire is okay, continue checking by connecting the probe to other leads. Any indication of resistance indicates that the wire is shorted to one of the other wires and that the harness is defective.

Ammeter

An ammeter measures current flow in a circuit. The ammeter must be placed into the circuit or in series with the circuit being tested. Normally, this requires disconnecting a wire or connector from a component and connecting the ammeter between the wire or connector and the component. The red lead of the ammeter should always be connected to the side of the connector closest to the positive side of the battery and the black lead should be connected to the other side.

It is much easier to test current using an ammeter with an inductive pickup. The pickup clamps around the wire or cable being tested. These ammeters measure amperage based on the magnetic field created by the current flowing through the wire. This type of pickup eliminates the need to separate the circuit to insert the meter.

Because ammeters are built with very low internal resistance, connecting them in series does not add any appreciable resistance to the circuit. Therefore, an accurate measurement of the current flow can be taken.

Volt/Ampere Tester

A volt/ampere tester (VAT) is used to test batteries, starting systems, and charging systems (Figure 2). The tester contains a voltmeter, ammeter, and carbon pile. The carbon pile is a variable resistor. A knob on the tester allows the technician to vary the resistance of the pile. When the tester is attached to the battery, the carbon pile will draw current out of the battery. The ammeter will read the amount of current draw. When testing a bat-

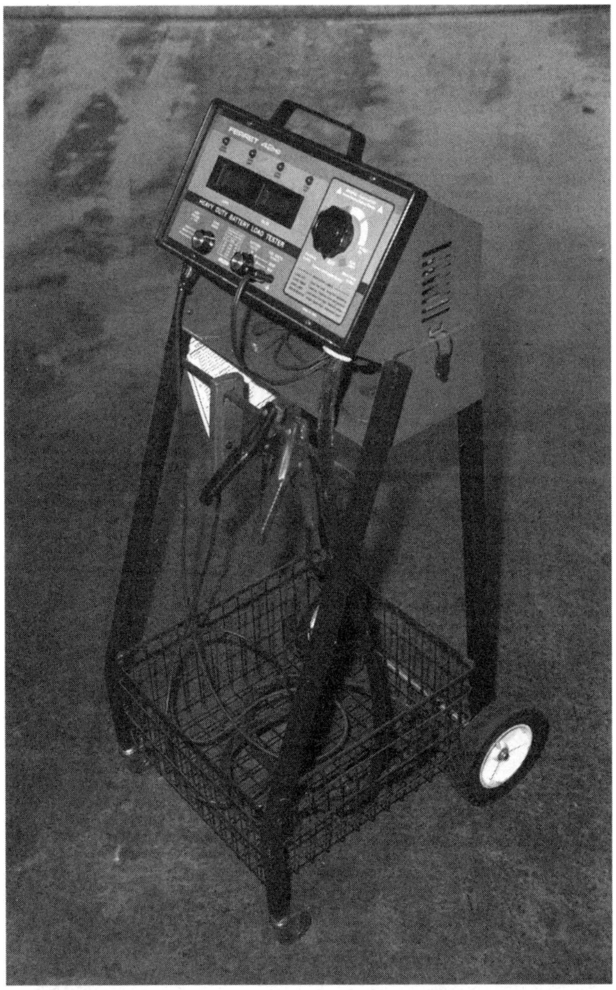

Figure 2 A VAT used to test batteries and starting and charging systems.

tery, the resistance of the carbon pile must be adjusted to match the ratings of the battery.

Conductance Tester

A conductance tester is a battery tester. Conductance is a measurement of the battery's ability to produce current. To measure conductance, the tester creates an AC voltage of a known frequency and amplitude that is sent through the battery, and then measures a portion of the AC current response. The degree of conductance is based on the plate surface that is available in the battery, which determines how much power the battery can supply.

As a battery ages, the plate surface can sulfate or shed active material, which adversely affects its ability to perform. In addition, conductance can be used to detect cell defects, shorts, and open circuits, which will reduce the ability of the battery to deliver current. Using conductance, the testers are able to provide an accurate assessment of the battery's condition.

To use this tester, connect the two test leads to the positive and negative posts of the cell or battery being tested, a measurement is taken in a matter of seconds. A conductance measurement is displayed in Mhos or Siemens, sometimes abbreviated with a "G".

Logic Probes

In some circuits, pulsed or digital signals pass through the wires. These "on-off" digital signals either carry information or provide power to drive a component. Many sensors used in a computer-control circuit send digital information back to the computer. To check the continuity of the wires that carry digital signals, a logic probe can be used.

A logic probe (Figure 3) has three different colored LEDs. A red LED lights when there is high voltage at the point being probed. A green LED lights to indicate low voltage. And a yellow LED indicates the presence of a voltage pulse. The logic probe is powered by the circuit and reflects only the activity at the point being probed. When the probe's test leads are attached to a circuit, the LEDs display the activity.

If a digital signal is present, the yellow LED will turn on. When there is no signal, the LED is off. If voltage is present, the red or green LEDs will light, depending on the amount of voltage. When there is a digital signal and the voltage cycles from low to high, the yellow LED will be lit and the red and green LEDs will cycle, indicating a change in the voltage.

Figure 3 A logic probe.

DMMs

It's not necessary for a technician to own separate meters to measure volts, ohms, and amps; a multimeter can be used instead. Top-of-the-line multimeters are multifunctional. Most test volts, ohms, and amperes in both DC and AC. Usually there are several test ranges provided for each of these functions. In addition to these basic electrical tests, multimeters also test engine rpm, duty cycle, pulse width, diode condition, frequency, and even temperature. The technician selects the desired test range by turning a control knob on the front of the meter.

Multimeters are available with either analog or digital displays, but the most commonly used multimeter is the digital volt/ohmmeter (DVOM), which is often referred to as a digital multimeter (DMM). There are several drawbacks to using analog-type meters for testing electronic control systems. Many electronic components require very precise test results. Digital meters can measure volts, ohms, or amperes in tenths and hundredths. Another problem with analog meters is their low internal resistance (input impedance). The low input impedance allows too much current to flow through circuits and should not be used on delicate electronic devices.

Digital meters, on the other hand, have a high input impedance, usually at least 10 megohms (10 million ohms). Metered voltage for resistance tests is well below 5 volts, reducing the risk of damage to sensitive components and delicate computer circuits. A high-impedance digital multimeter must be used to test the voltage of some components and systems such as an oxygen (O_2) sensor circuit. If a low-impedance analog meter is used in this type of circuit, the current flow through the meter is high enough to damage the sensor.

DMMs have either an "auto range" feature, in which the appropriate scale is automatically selected by the meter, or they must be set to a particular range. In either case, you should be familiar with the ranges and the different settings available on the meter you are using. To designate particular ranges and readings, meters display a prefix before the reading or range. If the meter has a setting for mAmps, that means the readings will be given in milli-amps or 1/1000th of an amp. Ohmmeter scales are expressed as a multiple of tens or use the prefix K or M. K stands for Kilo or 1000. A reading of 10K ohms equals 10,000 ohms. An M stands for Mega or 1,000,000. A reading of 10M ohms equals 10,000,000 ohms. When using a meter with an auto range, make sure you note the range being used by the meter. There is a big difference between 10 ohms and 10,000,000 ohms.

After the test range has been selected, the meter is connected to the circuit in the same way as if it were an individual meter.

When using the ohmmeter function, the DMM will show a zero or close to zero when there is good continuity. If the continuity is very poor, the meter will display an infinite reading. This reading is usually shown as a blinking "1.000", a blinking "1", or an "OL" (Figure 4). Before taking any measurement, calibrate the meter. This is done by holding the two leads together and adjusting the meter reading to zero. Not all meters need to be calibrated; some digital meters automatically calibrate when a scale is selected. On meters that require calibration, it is recommended that the meter be zeroed after changing scales.

Multimeters may also have the ability to measure duty cycle, pulse width, and frequency. All of these represent voltage pulses caused by the turning on and off of a circuit or the increase and

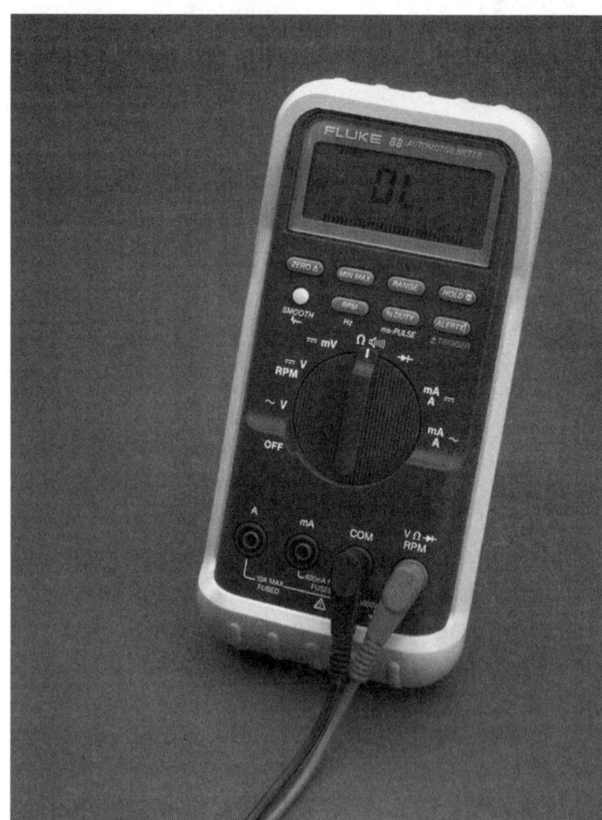

Figure 4 The infinite reading on a DMM.

decrease of voltage in a circuit. Duty cycle is a measurement of the amount of time something is on compared to the time of one cycle and is measured in a percentage.

Pulse width is similar to duty cycle except that it is the exact time something is turned on and is measured in milliseconds. When measuring duty cycle, you are looking at the amount of time something is on during one cycle.

The number of cycles that occur in one second is called the frequency. The higher the frequency, the more cycles occur in a second. Frequencies are measured in Hertz. One Hertz is equal to one cycle per second.

Lab Scopes

An oscilloscope is a visual voltmeter. An oscilloscope converts electrical signals to a visual image representing voltage changes over a specific period of time. This information is displayed in the form of a continuous voltage line called a waveform pattern or trace.

An oscilloscope screen is a cathode ray tube (CRT), which is very similar to the picture tube in a television set. High voltage from an internal source is supplied to an electron gun in the back of the CRT when the oscilloscope is turned on. This electron gun emits a continual beam of electrons against the front of the CRT. The external leads on the oscilloscope are connected to deflection plates above and below and on each side of the electron beam. When a voltage signal is supplied from the external leads to the deflection plates, the electron beam is distorted and strikes the front of the screen in a different location to indicate the voltage signal from the external leads.

An upward movement of the voltage trace on an oscilloscope screen indicates an increase in voltage, and a downward movement of this trace represents a decrease in voltage. As the voltage trace moves across an oscilloscope screen, it represents a specific length of time.

The size and clarity of the displayed waveform is dependent on the voltage scale and the time reference selected. Most scopes are equipped with controls that allow voltage and time interval selection. It is important, when choosing the scales, to remember that a scope displays voltage over time.

Dual-trace oscilloscopes can display two different waveform patterns at the same time (Figure 5). This makes cause and effect analysis easier.

With a scope, precise measurement is possible. A scope will display any change in voltage as it occurs. This is especially important for diagnosing intermittent problems.

The screen of a lab scope is divided into small divisions of time and voltage. Time is represented by the horizontal movement of the waveform. Voltage is measured with the vertical position of the waveform. Since the scope displays voltage over time, the waveform moves from the left (the beginning of measured time) to the right (the end of measured time). The value of the divisions can be adjusted to improve the view of the voltage waveform.

Since a scope displays actual voltage, it will display any electrical noise or disturbances that accompany the voltage signal. Noise is primarily caused by radio frequency interference (RFI), which may come from the ignition system. RFI is an unwanted voltage signal that rides on a signal. This noise can cause intermittent problems with unpredictable results. The noise causes slight increases and decreases in the voltage. When a computer receives a voltage signal with noise, it will try to react to the minute changes. As a result, the computer responds to the noise rather than the voltage signal.

Scan Tools

The introduction of computer-controlled systems brought with it the need for tools capable of troubleshooting electronic control systems. There are a variety of computer scan tools available today

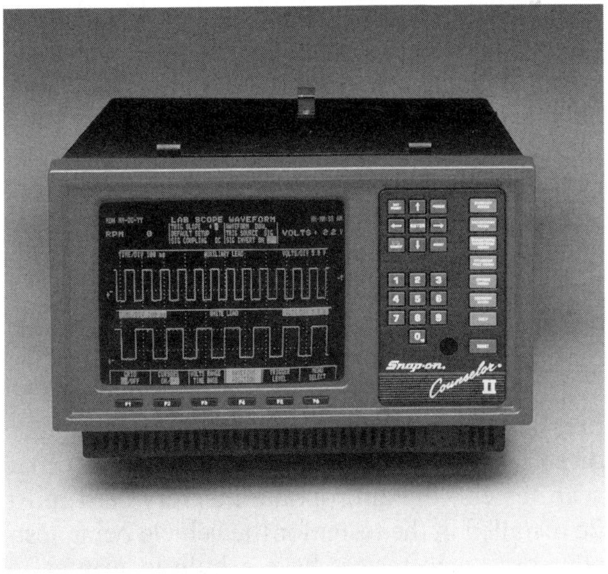

Figure 5 A dual-trace lab scope. *Courtesy of Snap-On Tool Company.*

that do just that. A scan tool is a microprocessor designed to communicate with the vehicle's computer. Connected to the computer through diagnostic connectors, a scan tool can access trouble codes, run tests to check system operations, and monitor the activity of the system. Trouble codes and test results are displayed on an LED screen, or printed out on the scanner printer.

Scan tools retrieve fault codes from a computer's memory and digitally display these codes on the tool. A scan tool may also perform many other diagnostic functions depending on the year and make of the vehicle. Most aftermarket scan tools have removable modules that are updated each year. These modules are designed to test the computer systems on various makes of vehicles. For example, some scan testers have a 3-in-1 module that tests the computer systems on Chrysler, Ford, and General Motors vehicles. A 10-in-1 module is also available to diagnose computer systems on vehicles imported by 10 different manufacturers. These modules plug into the scan tool.

Scan tools are capable of testing many onboard computer systems, such as climate controls, transmission controls, engine computers, antilock brake computers, air bag computers, and suspension computers, depending on the year and make of the vehicle and the type of scan tester. In many cases, the technician must select the computer system to be tested with the scanner after it has been connected to the vehicle.

The scan tool is connected to specific diagnostic connectors on various vehicles. Most manufacturers have one diagnostic connector that connects the data wire from each onboard computer to a specific terminal in the connector. Other vehicle manufacturers have several different diagnostic connectors on each vehicle, and each of these connectors may be connected to one or more onboard computers. A set of connectors is supplied with the scanner to allow tester connection to various diagnostic connectors on different vehicles.

The scanner must be programmed for the model year, make of vehicle, and type of engine. With some scan tools, this selection is made by pressing the appropriate buttons on the tester, as directed by the digital tester display. On other scan testers, the appropriate memory card must be installed in the tester for the vehicle being tested. Some scan testers have a built-in printer to print test results, while other scan testers may be connected to an external printer.

As automotive computer systems become more complex, the diagnostic capabilities of scan testers continue to expand. Many scan testers now have the capability to store, or "freeze," data into the tester during a road test, and then play back the data when the vehicle is returned to the shop.

Some scan testers now display diagnostic information based on the fault code in the computer memory. Service bulletins published by the manufacturer of the scan tester may be indexed by the tester after the vehicle information is entered in the tester. Other scan testers display sensor specifications for the vehicle being tested.

The vehicle's computer sets trouble codes when a voltage signal is entirely out of its normal range. The codes help technicians identify the cause of the problem when this is the case. If a signal is within its normal range but is still not correct, the vehicle's computer will not display a trouble code. However, a problem may still exist.

With OBD-II, the diagnostic connectors are located in the same place on all vehicles. Also, any scan tools designed for OBD-II will work on all OBD-II systems, therefore the need to have designated scan tools or cartridges is eliminated. The OBD-II scan tool has the ability to run diagnostic tests on all systems and has "freeze frame" capabilities.

Stethoscope

Some sounds can be easily heard without using a listening device, but others are impossible to hear unless they are amplified. A stethoscope is very helpful in locating the cause of a noise by amplifying the sound waves. It can also help you distinguish between normal and abnormal noise. The procedure for using a stethoscope is simple. Use the metal prod to trace the sound until it reaches its maximum intensity. Once the precise location has been discovered, the sound can be better evaluated. A sounding stick, which is nothing more than a long, hollow tube, works on the same principle, though a stethoscope gives much clearer results.

The best results, however, are obtained with an electronic listening device. With this tool you can tune into the noise. Doing this allows you to eliminate all other noises that might distract or mislead you.

Belt Tension Gauge

A belt tension gauge is used to measure drive belt tension. The belt tension gauge is installed over the belt, and the gauge indicates the amount of belt tension.

Vacuum Gauge

Often vacuum devices are tied into electrical circuits. Before you can thoroughly diagnose these systems, you must know the engine's ability to form a vacuum. Vacuum tests do just that. Manifold vacuum is tested with a vacuum gauge. Vacuum is formed on a piston's intake stroke. As the piston moves down, it lowers the pressure of the air in the cylinder—if the cylinder is sealed. This lower cylinder pressure is called engine vacuum. If there is a leak, atmospheric pressure will force air into the cylinder and the resultant pressure will not be as low. The reason atmospheric pressure enters is simply that whenever there is a low and high pressure, the high pressure will always move toward the low pressure.

Vacuum is measured in inches of mercury (in./Hg) and in kilopascals (kPa) or millimeters of mercury (mm/Hg).

To measure vacuum, a flexible hose on the vacuum gauge is connected to a source of manifold vacuum, either on the manifold or at a point below the throttle plates. Sometimes this requires removing a plug from the manifold and installing a special fitting.

The test is made with the engine cranking or running. A good vacuum reading is typically at least 16 in./Hg. However, a reading of 15 to 20 in./Hg (50 to 65 kPa) is normally acceptable. Since the intake stroke of each cylinder occurs at a different time, the production of vacuum occurs in pulses. If the amount of vacuum produced by each cylinder is the same, the vacuum gauge will show a steady reading. If one or more cylinders are producing different amounts of vacuum, the gauge will show a fluctuating reading.

Oil Pressure Gauge

When an oil pressure gauge is suspected of giving an erroneous reading, it may be necessary to check the oil pressure of the engine. Oil pressure is checked at the sending unit passage with an externally mounted mechanical oil pressure gauge. Various fittings are usually supplied with the oil pressure gauge to fit different openings in the lubrication system.

Battery Hydrometer

On unsealed batteries, the specific gravity of the electrolyte can be measured to give a fairly good indication of the battery's state of charge. A hydrometer is used to perform this test. A battery hydrometer (Figure 6) consists of a glass tube or barrel, rubber bulb, rubber tube, and a glass float or hydrometer with a scale built into its upper stem. The glass tube encases the float and forms a reservoir for the test electrolyte. Squeezing the bulb pulls electrolyte into the reservoir.

When filled with test electrolyte, the sealed hydrometer float bobs in the electrolyte. The depth to which the glass float sinks in the test electrolyte indicates its relative weight compared to water. The reading is taken off the scale by sighting along the level of the electrolyte.

If the hydrometer floats deep in the electrolyte, the specific gravity is low. If the hydrometer floats shallow in the electrolyte, the specific gravity is high.

At extremely high and low electrolyte temperatures, it is necessary to correct the reading by adding or subtracting 4 points (0.004) for each 10°F above or below the standard of 80°F. Most hydrometers have a built-in thermometer to measure the temperature of the electrolyte. The hydrometer reading can be misleading if it is not adjusted. It is important to make these adjustments at high and low temperatures to determine the battery's true state of charge.

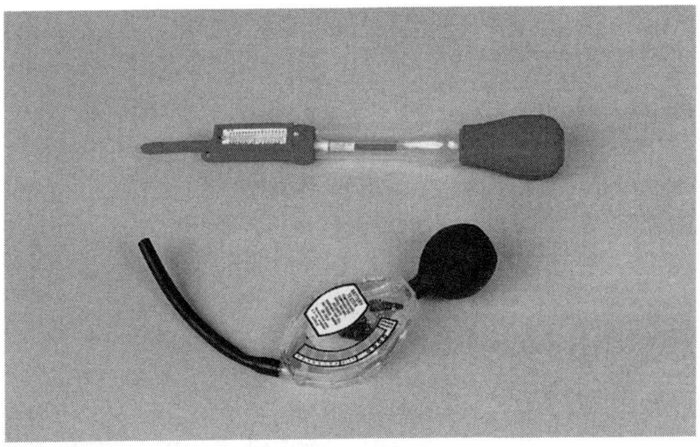

Figure 6 Battery hydrometers.

On many sealed maintenance-free batteries a special temperature-compensated hydrometer is built into the battery cover. A quick visual check indicates the battery's state of charge. The hydrometer has a green ball within a cage that is attached to a clear plastic rod. The green ball floats at a predetermined specific gravity of the electrolyte that represents about a 65 percent state of charge. When the green ball floats, it rises within the cage and positions itself under the rod. Visually, a green dot then shows in the center of the hydrometer. In testing, the green dot means the battery is charged enough for testing. If the green dot is not visible and has a dark appearance, it means the battery must be charged before the test procedure is performed.

While charging, the appearance of the green dot means that the battery is sufficiently charged. Charging can be stopped to prevent overcharging.

It is important when you are observing the hydrometer that the battery has a clean top so that you can see the correct indication. A flashlight may be required in some poorly lit areas. Always look straight down when viewing the hydrometer.

On some special applications, some hydrometers feature a red dot indication in addition to the green dot, dark, and clear appearances. The red dot means the battery is nearing complete discharge and must be charged before being used.

Wire and Terminal Repair Tools

Many automotive electrical problems can be traced to faulty wiring. Loose or corroded terminals, frayed, broken, or oil soaked wires, and faulty insulation are the most common causes.

Wire end terminals are connecting devices. They are generally made of tin-plated copper and come in many shapes and sizes. They may be either soldered or crimped in place. Common tools for making wiring repairs include a wire stripper (Figure 7), soldering iron or gun, and a crimping tool (Figure 8). When installing a terminal, always use the proper crimping tool and follow the tool manufacturer's instructions.

Headlight Aimers

Headlights must be kept in adjustment to obtain maximum illumination. Sealed beams that are properly adjusted cover the correct range and afford the driver the proper nighttime view. Headlights can be adjusted using headlamp adjusting tools (Figure 9) or by shining the lights on a chart. Headlight aiming tools give the best results with the least amount of work. It should be noted that many late-model vehicles have levels built into the headlamp assemblies. These are used to correctly adjust the headlights.

Most headlight aimers use mirrors with split images, similar to split-image finders on some cameras, and spirit levels to determine exact adjustment. When using any headlight aiming equipment, follow the instructions provided by the equipment manufacturer.

Feeler Gauge

A feeler gauge is a thin strip of metal or plastic of known and closely controlled thickness. Several of these strips are often assembled together as a feeler gauge set that looks like a pocketknife. A gauge of a desired thickness can be pivoted away from the others for convenient use. A feeler gauge set usually contains strips or leaves of 0.002- to 0.010-inch thickness (in steps of 0.001 inch) and leaves of 0.012- to 0.024-inch thickness (in steps of 0.002 inch).

A feeler gauge can be used by itself to measure clearances, such as critical compressor clutch clearances. It can also be used with a precision straightedge to measure the flatness of sealing surfaces in a compressor.

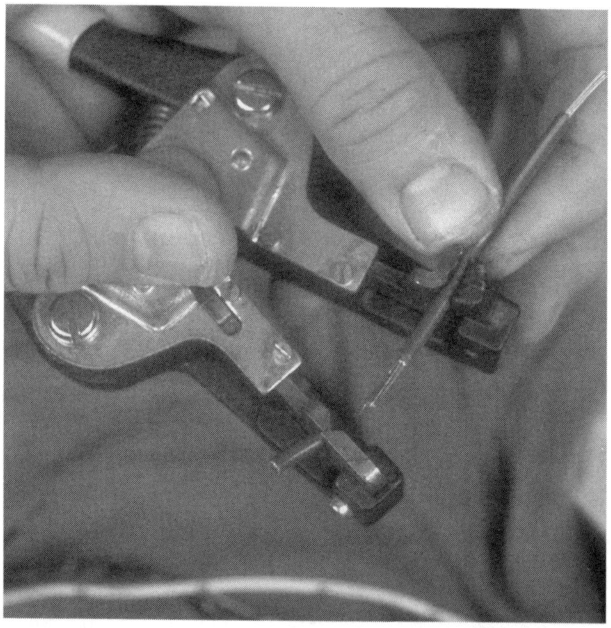

Figure 7 A wire-stripping tool.

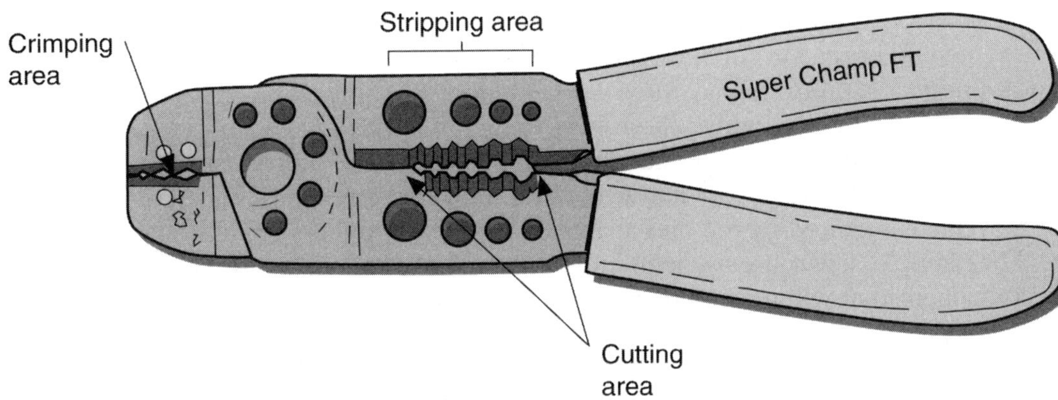

Figure 8 A wire terminal crimping tool.

Straightedge

A straightedge is no more than a flat bar machined to be totally flat and straight and, to be effective, it must be flat and straight. Any surface that should be flat can be checked with a straightedge and feeler gauge set. The straightedge is placed across and at angles on the surface. At any low points on the surface, a feeler gauge can be placed between the straightedge and the surface. The size of the gauge that fills in the gap indicates the amount of warpage or distortion.

Dial Indicator

The dial indicator is calibrated in 0.001-inch (one-thousandth inch) increments. Metric dial indicators are also available. Both types are used to measure movement. Common uses of the dial indicator include checking compressor shaft runout, shaft endplay, and compressor clutch runout.

To use a dial indicator, position the indicator rod against the object to be measured. Then, push the indicator toward the work until the indicator

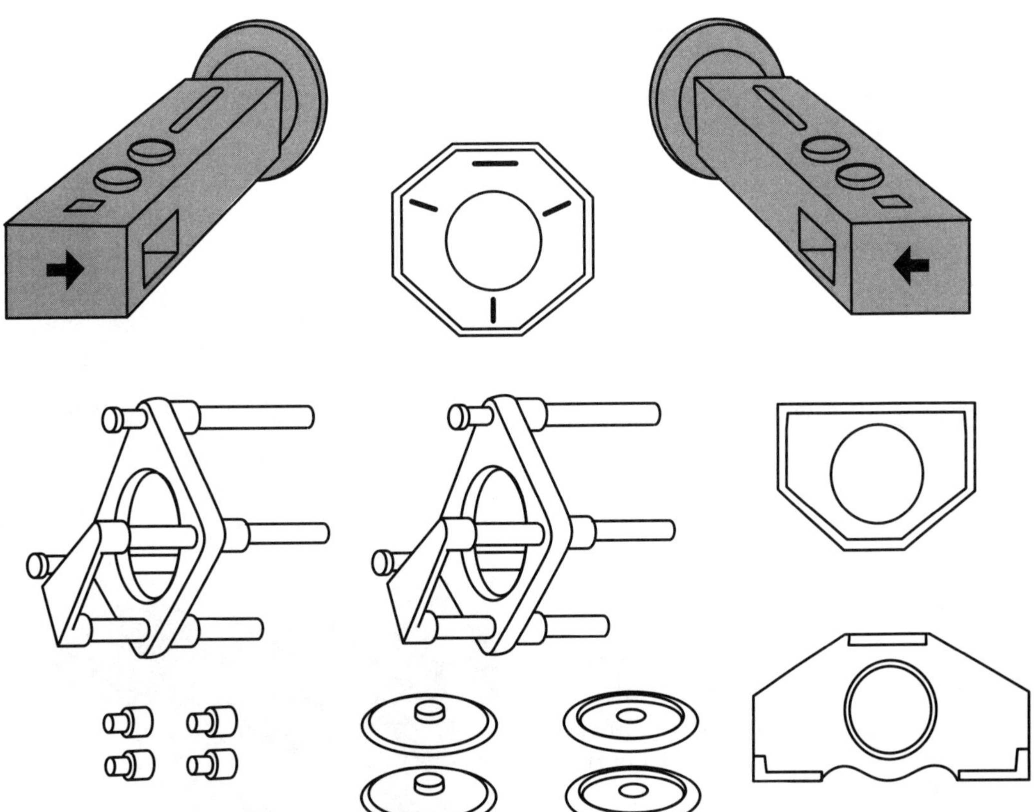

Figure 9 Headlight aiming kit. *Courtesy of DaimlerChrysler Corporation.*

needle travels far enough around the gauge face to permit movement to be read in either direction. Zero the indicator needle on the gauge. Move the object in the direction required, while observing the needle of the gauge. Always be sure the range of the dial indicator is sufficient to allow the amount of movement required by the measuring procedure. For example, never use a 1-inch indicator on a component that will move 2 inches.

Torque-Indicating Wrench

Torque is the twisting force used to turn a fastener against the friction between the threads and between the head of the fastener and the surface of the component. The fact that practically every vehicle and engine manufacturer publishes a list of torque recommendations is ample proof of the importance of using proper amounts of torque when tightening nuts or bolts. The amount of torque applied to a fastener is measured with a torque-indicating or torque wrench.

There are three basic types of torque-indicating wrenches available with pounds per inch and pounds per foot increments: a beam torque wrench that has a beam that points to the torque reading, a "click"-type torque wrench in which the desired torque reading is set on the handle (when the torque reaches that level, the wrench clicks), and a dial torque wrench that has a dial that indicates the torque exerted on the wrench. Some designs of this type torque wrench have a light or buzzer that turns on when the desired torque is reached.

Gear and Bearing Pullers

Many tools are designed for a specific purpose. An example of a special tool is a gear and bearing puller. Many gears and bearings have a slight interference fit (press-fit) when they are installed on a shaft or in a housing. Something that has a press-fit has an interference fit. For example, if the inside diameter of a bore is 0.001 inch smaller than the outside diameter of a shaft, when the shaft is fitted into the bore it must be pressed in to overcome the 0.001 inch interference fit. This press-fit prevents the parts from moving on each other. The removal of these gears and bearings must be done carefully to prevent damage to the gears, bearings, or shafts. Prying or hammering can break or bind the parts. A puller with the proper jaws and adapters should be used to remove gears and bearings. Using the proper puller, the force required to remove a gear or bearing can be applied with a slight and steady motion.

There are pullers specifically designed for removing battery terminals from a battery (Figure 10). These pullers should be used whenever the terminal doesn't slip off. Prying on the terminal to remove it can damage the battery.

Bushing and Seal Pullers and Drivers

Another commonly used group of special tools are the various designs of bushing and seal drivers and pullers. Pullers are either a threaded or slide hammer type tool. Always make sure you use the correct tool for the job; bushings and seals are easily damaged if the wrong tool or procedure is used.

Retaining Ring Pliers

A technician will run into many different styles and sizes of retaining rings and snap rings that hold things together or keep them in a fixed location. Using the correct tool to remove and install these rings is the only safe way to work with them. All technicians should have an assortment of snap ring and retaining ring pliers.

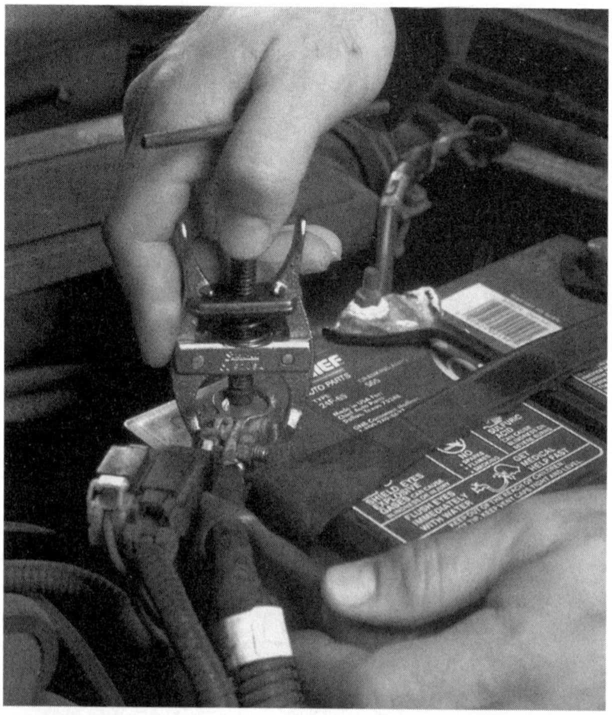

Figure 10 A battery terminal puller.

Service Manuals

Perhaps the most important tools you will use are service manuals. There is no way a technician can remember all of the procedures and specifications needed to correctly repair all vehicles. Therefore, a good technician relies on service manuals and other information sources for this information. Good information plus knowledge allows a technician to fix a problem with the least bit of frustration and at the lowest cost to the customer.

To obtain the correct system specifications and other information, you must first identify the exact system you are working on. The best source for vehicle identification is the VIN. The code can be interpreted through information given in the service manual. The manual may also help you identify the system through identification of key components or other identification numbers and/or markings.

The primary source of repair and specification information for any car, van, or truck is the manufacturer. The manufacturer publishes service manuals each year, for every vehicle built. Because of the enormous amount of information, some manufacturers publish more than one manual per year per car model. The manuals are typically divided into sections based on the major systems of the vehicle. In the case of air conditioning systems, there is a section for each engine that may be in the vehicle. Often the heating system is included with information about the engine's cooling system. Manufacturers' manuals cover all repairs, adjustments, specifications, detailed diagnostic procedures, and the special tools required.

Since many technical changes occur on specific vehicles each year, manufacturers' service manuals need to be constantly updated. Updates are published as service bulletins (often referred to as Technical Service Bulletins or TSBs) that show the changes in specifications and repair procedures during the model year. These changes do not appear in the service manual until the next year. The car manufacturer provides these bulletins to dealers and repair facilities on a regular basis.

Service manuals are also published by independent companies rather than the manufacturers. However, they pay for and get most of their information from the car makers. These manuals contain component information, diagnostic steps, repair procedures, and specifications for several car makes in one book. Information is usually condensed and is more general in nature than the manufacturer's manuals. The condensed format allows for more coverage in less space and, therefore, is not always specific. They may also contain several years of models as well as several car makes in one book.

Many of the larger parts manufacturers have excellent guides on the various parts they manufacture or supply. They also provide updated service bulletins on their products. Other sources for up-to-date technical information are trade magazines and trade associations.

The same information that is available in service manuals is now commonly found electronically on compact disks (CD-ROMs), digital video disks (DVDs), and the Internet. A single compact disk can hold a quarter million pages of text, eliminating the need for a huge library to contain all of the printed manuals. Using electronics to find information is also easier and quicker. The disks are normally updated quarterly and not only contain the most recent service bulletins but also engineering and field service fixes. DVDs can hold more information than CDs; therefore, fewer disks are needed with systems that use DVDs. The CDs and DVDs are inserted into a computer. All a technician needs to do is enter vehicle information and then move to the appropriate part or system. The appropriate information will then appear on the computer's screen. Online data can be updated instantly and requires no space for physical storage. These systems are easy to use and the information is quickly accessed and displayed. The computer's keyword, mouse, and/or light pen are used to make selections from the screen's menu. Once the information is retrieved, a technician can read it off the screen or print it out and take it to the service bay.

CROSS-REFERENCE GUIDE

NATEF Task	Job Sheet
A.1	1
A.2	2
A.3	3
A.4	3
A.5	4
A.6	5
A.7	6
A.8	7
A.9	8
A.10	9
A.11	10
A.12	11
A.13	12
A.14	9
A.15	13
A.16	11
A.17	14
A.18	14
A.19	14
A.20	15
A.21	16
B.1	17
B.2	17
B.3	18
B.4	19
B.5	20
B.6	19
B.7	21
B.8	16
B.9	22
B.10	23
C.1	24 (Job sheet 26 is related)
C.2	24
C.3	24
C.4	25 (Job sheet 28 is related)
C.5	27
C.6	29
D.1	30

D.2	30 (Job sheet 32 is related)
D.3	31
D.4	33 (Job sheet 34 is related)
D.5	30
E.1	35
E.2	36
E.3	35
E.4	37
F.1	38
F.2	39
F.3	40
F.4	41
G.1	42
G.2	43
G.3	44
H.1	45
H.2	46
H.3	47
H.4	48
H.5	49
H.6	50
H.7	51
H.8	52
H.9	53
H.10	53
H.11	54

JOB SHEETS

ELECTRICAL AND ELECTRONIC SYSTEMS JOB SHEET 1

Filling Out a Work Order

Name ______________________ Station ______________ Date ______________

NATEF Correlation:

This Job Sheet addresses the following NATEF task:

A.1. Complete work order to include customer information, vehicle identifying information, customer concern, related service history, cause, and correction.

Objective

Upon completion of this job sheet, you will be able to prepare a service work order based on customer input, vehicle information, and service history.

Tools and Materials

An assigned vehicle or the vehicle of your choice

Service work order or computer-based shop management package

Parts and labor guide

Work Order Source: Describe the system used to complete the work order. If a paper repair order is being used, describe the source.

____

____

PROCEDURE

1. Prepare the shop management software for entering a new work order or obtain a blank paper work order. Task Completed ☐
2. Enter customer information, including name, address, and phone numbers onto the work order. Task Completed ☐
3. Locate and record the vehicle's VIN. Task Completed ☐
4. Enter the necessary vehicle information, including year, make, model, engine type and size, transmission type, license number, and odometer reading. Task Completed ☐
5. Does the VIN verify that the information about the vehicle is correct?

   ____
6. Normally, you would interview the customer to identify his or her concerns. However, to complete this job sheet, assume the only concern is that the right front headlamp does not work when the high beams are selected. This concern should be added to the work order. Task Completed ☐

7. The history of service to the vehicle can often help diagnose problems as well as indicate possible premature part failure. Gathering this information from the customer can provide some of this information. For this job sheet assume the vehicle has not had a similar problem and was not recently involved in a collision. Service history is further obtained by searching files for previous service. Often this search is done by customer name, VIN, and license number. Check the files for any related service work. Task Completed ☐

8. Search for technical service bulletins on this vehicle that may relate to the customer's concern. Task Completed ☐

9. Based on the customer's concern, service history, TSBs, and your knowledge, what is the likely cause of this concern?

____

____

10. Enter this information onto the work order. Task Completed ☐

11. Prepare to make a repair cost estimate for the customer. Identify all parts that may need to be replaced to correct the concern. List these here.

____

____

12. Describe the task(s) that will be necessary to replace the part.

____

____

13. Using the parts and labor guide, locate the cost of the parts that will be replaced and enter the cost of each item onto the work order at the appropriate place for creating an estimate. Task Completed ☐

14. Now, locate the flat rate time for work required to correct the concern. List each task and with its flat rate time.

____

____

15. Multiply the time for each task by the shop's hourly rate and enter the cost of each item onto the work order at the appropriate place for creating an estimate. Task Completed ☐

16. Many shops have a standard amount they charge each customer for shop supplies and waste disposal. For this job sheet, use an amount of ten dollars for shop supplies. Task Completed ☐

17. Add the total costs and insert the sum as the subtotal of the estimate. Task Completed ☐

18. Taxes must be included in the estimate. What is the sales tax rate and does it apply to both parts and labor, or just one of these?

____

____

19. Enter the appropriate amount of taxes to the estimate, then add this to the subtotal. The end result is the estimate to give the customer. Task Completed ☐

20. By law, how accurate must your estimate be?

____

____

21. Generally speaking, the work order is complete and is ready for the customer's signature. However, some businesses require additional information; make sure you enter that information to the work order. On the work order there is a legal statement that defines what the customer is agreeing to. Briefly describe the contents of that statement.

____

____

Problems Encountered

____

____

____

Instructor's Comments

____

____

____

ELECTRICAL AND ELECTRONIC SYSTEMS JOB SHEET 2

Identifying Problems and Concerns

Name ______________________ Station ______________ Date ______________

NATEF Correlation

This Job Sheet addresses the following NATEF task:

A.2. Identify and interpret electrical/electronic system concern; determine necessary action.

Objective

Upon completion of this job sheet, you will be able to define electrical and electronic system problems or concerns, prior to diagnosing or testing the systems.

Protective Clothing

Goggles or safety glasses with side shields

Describe the vehicle being worked on:

Year ________ Make ____________________ Model ____________________

VIN ____________________ Engine type and size ____________________

PROCEDURE

1. Conduct a visual inspection of the fuses, obvious wiring and connectors, battery, and generator drive belt. Describe your findings.

   ____

   ____

   ____

   ____

2. What would be indicated by a blown fuse and how would you locate the cause for the bad fuse?

   ____

   ____

3. Start the engine and describe how the starter system seemed to work. Include the speed of the starter motor and all noises.

   ____

   ____

4. Based on starter operation, what can you assume about the vehicle's battery and starting system?

____

____

5. Position the vehicle so the light from the headlamps are easily observed. Then with the engine running and the transmission in park or neutral and the parking brake applied, turn on the headlamps. Look for any change in brightness as you increase and decrease the engine's speed. Describe what you observed.

____

____

6. Based on the above check what do you know about the charging system, battery, and headlamps?

____

____

7. Check all of the lights in and on the vehicle. Describe your findings.

____

____

8. What basic type of problem would be indicated if one bulb did not light when it should have?

____

____

9. What basic type of problem would be indicated if one or more bulbs were not as bright as the others in the circuit?

____

____

10. Check all of the electrical accessories in and on the vehicle for operation and describe any problems. If you find anything that works abnormally or doesn't work, describe the type of problem you suspect is causing the malfunction.

____

____

____

Problems Encountered

____

____

____

Instructor's Comments

____

____

____

ELECTRICAL AND ELECTRONIC SYSTEMS JOB SHEET 3

Gathering Vehicle Information

Name ______________________ Station ______________ Date ______________

NATEF Correlation

This Job Sheet addresses the following NATEF tasks:

A.3. Research applicable vehicle and service information, such as electrical/electronic system operation, vehicle service history, service precautions, and technical service bulletins.

A.4. Locate and interpret vehicle and major component identification numbers (VIN, vehicle certification labels, calibration labels).

Objective

Upon completion of this job sheet, you will be able to gather service information about a vehicle and its electrical/electronic system.

Tools and Materials

Appropriate service manuals

Computer

Protective Clothing

Goggles or safety glasses with side shields

Describe the vehicle being worked on:

Year ________ Make ____________________ Model ____________________

VIN ____________________

PROCEDURE

1. Using the service manual or other information source, describe what each letter and number in the VIN for this vehicle represents.

____

____

2. Locate the Vehicle Emissions Control Information (VECI) label and describe where you found it.

____

____

3. Summarize what information you found on the VECI label.

____

____

4. While looking in the engine compartment did you find a calibration label? Describe where you found it.

____

____

5. Summarize the information contained on this label. Describe how you may use this information.

____

____

____

____

6. Look closely at the generator for all numbers stamped in the housing or for any labeling that identifies the rating of the generator. Record your findings here.

____

____

____

7. Using the information sources you have available, describe the generator.

____

____

____

8. Locate the labeling on the battery and describe the capacity rating and age of the battery.

____

____

____

9. Using a service manual or electronic database, locate the information about the vehicle's electrical and electronic system. List the major electronic components of the system and describe how the system is controlled.

____

____

____

____

____

10. Using a service manual or electronic database, locate and record all service precautions noted by the manufacturer regarding the electronic and electrical systems.

11. Using the information that is available, locate and record the vehicle's service history.

12. Using the information sources that are available, summarize all Technical Service Bulletins for this vehicle that relate to the electrical and electronic systems.

Problems Encountered

Instructor's Comments

ELECTRICAL AND ELECTRONIC SYSTEMS JOB SHEET 4

Using Ohm's Law During Diagnosis

Name ______________________ Station ______________ Date ______________

NATEF Correlation

This Job Sheet addresses the following NATEF task:

A.5. Diagnose electrical/electronic integrity for series, parallel, and series-parallel circuits using principles of electricity (Ohm's Law).

Objective

Upon completion of this job sheet, you will be able to use Ohm's law to help you logically diagnose electrical and electronic problems.

Tools and Materials

Paper and pen

PROCEDURE

NOTE: *Studying Ohm's Law is not merely an exhaustive exercise with numbers and formulas. Ohm's Law describes how electrical circuits behave, and by knowing how a circuit should behave, it is easier to determine if something is wrong and what the basic problem is. This worksheet is not to be completed on a vehicle, rather it should be completed with you sitting down and thinking and applying your understanding of the principles of electricity to answer these questions.*

1. Briefly state what Ohm's Law says. Do not enter the basic formula here.

____

____

2. Based on Ohm's Law, what is the basic formula for finding amperage (current) when you know the voltage and resistance of the circuit?

____

____

3. What is the basic difference between a series circuit and a parallel circuit?

____

____

____

____

4. What is different about a series-parallel circuit? Give an example.

____

____

____

____

5. If the circuit resistance in a series circuit is increased, what happens to circuit current?

____

____

6. If the circuit resistance in a parallel circuit is increased, what happens to circuit current?

____

____

7. If the resistance of one of the resistors or loads in a series circuit increases, what happens to circuit current?

____

____

8. If the resistance of one of the resistors or loads in a parallel circuit increases, what happens to circuit current?

____

____

9. If there is a corroded ground in a circuit, how are the circuit current and the voltage drops across the normal loads affected?

____

____

10. In a one-lamp light bulb circuit, if the bulb is burned out, what happens to the current in the circuit and how many volts would be dropped by the bulb?

____

____

11. What would happen to the operation of the circuit, as well as to the circuit current, if one leg (branch) of a parallel circuit is shorted to another leg of that circuit?

____

____

12. If one leg of a parallel circuit is shorted to ground, what happens to the other legs in that circuit? (Think deeply about this and talk about what happens before the circuit protection device opens!)

____

____

____

____

13. Fill-in-the-blanks

Based on Ohm's Law and observation, we know that whenever there is an increase in circuit current, the resistance must have __________. This could only be caused by a(n) __________ because a(n) __________ would result in no current and a(n) __________ problem would result in lower circuit current. We also know that when a light bulb burns dimmer than it should or a motor operates slower than it should, the cause must be __________ or lower source voltage. Further, we know if something doesn't work at all, the problem must be a(n) __________ or a(n) __________. If it is the latter, we will find an open circuit protection device or a burned wire.

Problems Encountered

____

____

____

Instructor's Comments

____

____

____

ELECTRICAL AND ELECTRONIC SYSTEMS JOB SHEET 5

Part Identification on a Wiring Diagram

Name ______________________ Station ______________ Date ______________

NATEF Correlation

This Job Sheet addresses the following NATEF task:

A.6. Use wiring diagrams during diagnosis of electrical circuit problems.

Objective

Upon completion of this job sheet, you will be able to identify different parts on a wiring diagram to enable the diagnosis of electrical circuit problems.

Tools and Materials

A wiring diagram in a service manual (assigned by the instructor)

Type of Vehicle:

Year ________ Make ____________________ Model ____________________

Service manual used: ______________________

Wiring diagram is found on pages: _______ to _______ .

PROCEDURE

Look over the wiring diagram, then answer the following questions:

1. Are all circuit grounds clearly marked in the wiring diagrams?

   ____

2. What is represented by lines that cross other lines?

   ____

   ____

3. Are most switches shown in their normally open or normally closed position?

   ____

4. Do all wires have a color code listed by them?

   ____

5. Is the internal circuitry of all components shown in their schematical drawing?

   ____

List the page and figure numbers (the location) where the following electrical components are shown in the wiring diagram. Then **draw the schematical symbol** used by this wiring diagram to represent the part.

Component	Location	Drawing
Windshield wiper motor	________________	________________
Dome (courtesy) light	________________	________________
A/C compressor clutch	________________	________________
Turn signal flasher unit	________________	________________
Fuse	________________	________________
Fuel gauge sending unit	________________	________________

Problems Encountered

____

____

____

Instructor's Comments

____

____

____

ELECTRICAL AND ELECTRONIC SYSTEMS JOB SHEET 6

Using a Digital Multimeter

Name ______________________ Station ______________ Date ______________

NATEF Correlation

This Job Sheet addresses the following NATEF task:

A.7. Demonstrate the proper use of a digital multimeter (DMM) during diagnosis of electrical circuit problems.

Objective

Upon completion of this job sheet, you will be able to use a digital multimeter for diagnosing electrical and electronic circuits.

Tools and Materials

Digital Multimeter (DMM)

PROCEDURE

1. List the manufacturer and model number of the DMM you are using for this worksheet.

   ____

   ____

2. Different DMMs are capable of measuring different things. What can this DMM do?

   ____

   ____

   ____

   ____

3. In order to use this DMM to full capacity do you need to insert a module or connect certain leads? If so, what are they?

   ____

   ____

4. According to the literature that came with the DMM, what are the special features of this DMM?

   ____

   ____

   ____

5. What is the impedance of the meter?

____

____

6. If the meter is auto-ranging, how many decimal points are shown on the meter and what suffixes are used?

____

____

____

____

7. Does the ohmmeter function of this meter need to be zeroed before use?

____

____

8. What are the meter's low and high limits for measuring voltage, amperage, and resistance?

____

____

Problems Encountered

____

____

____

Instructor's Comments

____

____

____

ELECTRICAL AND ELECTRONIC SYSTEMS JOB SHEET 7

Check Continuity in a Circuit with a Test Light

Name ______________________ Station ______________ Date ______________

NATEF Correlation

This Job Sheet addresses the following NATEF task:

A.8. Check electrical circuits with a test light; determine necessary action.

Objective

Upon completion of this job sheet, you will be able to test continuity in circuits with a test light.

WARNING: *Never use a regular test light on an electronic circuit; use only a high-impedance test light on these circuits.*

Tools and Materials

Test light

Wiring diagram

Service manual

Protective Clothing

Goggles or safety glasses with side shields

Describe the vehicle being worked on:

Year ________ Make ____________________ Model ____________________

VIN ____________________ Engine type and size ____________________

PROCEDURE

1. Determine if the circuit to be tested is a power supply circuit or a ground circuit. Task Completed ☐
2. If the circuit to be tested is a power supply circuit, connect the clip end of the test light to a ground on the vehicle. Task Completed ☐
3. Turn the key to the on position. Task Completed ☐
4. With the light end, use the probe to check the wire in the circuit at each connector in the specific circuit. Task Completed ☐
5. If the test light comes on, there is continuity between the power supply and the point where you are testing. Task Completed ☐
6. If the circuit to be tested is a ground circuit, connect the clip end of the test light to a power supply, such as the battery. Task Completed ☐
7. With the light end, use the probe to check the wire in the circuit at each connector in the specific circuit. Task Completed ☐

8. If the test light comes on, there is continuity between the ground and the point you are testing. Task Completed ☐

Problems Encountered

____

____

____

Instructor's Comments

____

____

____

ELECTRICAL AND ELECTRONIC SYSTEMS JOB SHEET 8

Check Voltage Drop Across Connectors

Name ______________________ Station ______________ Date ______________

NATEF Correlation

This Job Sheet addresses the following NATEF task:

A.9. Measure source voltage and perform voltage drop tests in electrical/electronic circuits using a voltmeter; determine necessary action.

Objective

Upon completion of this job sheet, you will be able to check voltage and voltage drop in electrical/electronic circuits using a digital multimeter.

Tools and Materials

Digital multimeter (DMM)

Wiring diagrams

Service manual

Protective Clothing

Goggles or safety glasses with side shields

Describe the vehicle being worked on:

Year ________ Make ____________________ Model ____________________

VIN ____________________ Engine type and size ____________________

PROCEDURE

1. Determine the connector to be tested. Task Completed ☐
2. Turn on the ignition to supply power to the circuit. Task Completed ☐
3. Set the DMM on the proper voltage scale. Task Completed ☐
4. Connect the negative lead (the black lead) of the meter to a ground on the vehicle. Task Completed ☐
5. Use the positive lead (the red lead) to check the voltage at one side of the connector. Task Completed ☐
6. Record the voltage. ________________________________
7. Use the positive lead (the red lead) to check the voltage at the other side of the connector. Task Completed ☐
8. Record the voltage. ________________________________

9. If there is a difference subtract the lower voltage from the higher voltage. Task Completed ☐

10. Compare this difference with the specifications in the service manual to determine if the difference is within specifications. Task Completed ☐

Problems Encountered

____

____

____

Instructor's Comments

____

____

____

ELECTRICAL AND ELECTRONIC SYSTEMS JOB SHEET 9

Using an Ammeter to Check Parasitic Battery Drain

Name ______________________________ Station __________________ Date __________________

NATEF Correlation

This Job Sheet addresses the following NATEF tasks:

A.10. Measure current flow in electrical/electronic circuits and components using an ammeter; determine necessary action.

A.14. Measure and diagnose the causes of abnormal key-off battery drain (parasitic draw); determine necessary action.

Objective

Upon completion of this job sheet, you will be able to check current flow in electrical and electronic circuits and components with an ammeter and measure and diagnose the causes of abnormal key-off battery drain.

Tools and Materials

Ammeter

Digital multimeter (DMM)

Protective Clothing

Goggles or safety glasses with side shields

Describe the vehicle being worked on:

Year __________ Make __________________________ Model __________________________

VIN __________________________ Engine type and size __________________________

PROCEDURE

1. If you are using a DMM to measure amperage, set the meter to the highest reading of amps provided by the meter. If using an ammeter, select the highest amperage scale available. Task Completed ☐
2. If the ammeter has an inductive pickup for current measurements, place the pickup clamp around the wire in the circuit being tested. In this case, put it around the negative battery cable. Task Completed ☐
3. If the meter does not have an inductive pickup, the meter must be connected in series with the circuit. Disconnect the negative cable of the battery. Connect the red meter lead to the cable end and the black meter lead to the battery. Task Completed ☐
4. With everything turned off and the doors closed, what is the amperage reading shown on the meter?

   ____

5. If there is no reading, select a lower amperage scale on the meter until you are able to take a reading. What was the reading?

____

____

6. Locate the allowable parasitic battery drain specification for the vehicle in the service manual. What is the maximum allowable amperage drain?

____

____

7. Compare your measurement to the specifications and give your conclusions about the test.

____

____

8. If the battery drain was excessive, visually check the trunk, glove box, and underhood lights to see if they are on. If any are on, remove the bulb and observe the ammeter. What did you observe?

____

____

9. If the drain is now within specifications, troubleshoot the removed light's circuit for a short. Task Completed ☐

10. If no lights were found to be on or the light was not the cause of excessive drain, go to the fuse panel or distribution center and remove one fuse at a time while watching the ammeter. Task Completed ☐

11. When the battery drain decreases, the circuit protected by the fuse you removed last is the source of the problem. Describe what happened.

____

____

12. Troubleshoot that circuit for a short and repair the problem. After the problem has been corrected, test the battery drain again to verify the fix. Task Completed ☐

Problems Encountered

____

____

____

Instructor's Comments

____

____

____

HALF PRICE BOOKS®

Half Price Books
1835 Forms Drive
Carrollton, TX 75006
OFS OrderID 39492466

Thank you for your order, Main Address!

Thank you for shopping with Half Price Books! Please contact Support@hpb.com. if you have any questions, comments or concerns about your order (113-6939673-7753045)

Visit our stores to sell your books, music, movies games for cash.

SKU	ISBN/UPC	Title & Author/Artist	Shelf ID	Qty	OrderSKU
S378507295	9781418020798	NATEF Standard Jobsheet A6 Erjavec, Jack	56--08--4	1	

SHIPPED STANDARD TO:
Main Address
800 Avondale Ave
1033837-1-0084
Grandview Heights OH 43212
j1gg094j97htty5@marketplace.amazon.com

ORDER# **113-6939673-7753045**
AmazonMarketplaceUS

ELECTRICAL AND ELECTRONIC SYSTEMS JOB SHEET 10

Check Continuity with an Ohmmeter

Name ______________________ Station ______________ Date ______________

NATEF Correlation

This Job Sheet addresses the following NATEF task:

A.11. Check continuity and resistances in electrical/electronic circuits and components with an ohmmeter; determine necessary action.

Objective

Upon completion of this job sheet, you will be able to check the continuity and resistances in electrical/electronic circuits and components with an ohmmeter.

Tools and Materials

A vehicle

Digital Multimeter (DMM)

An analog ohmmeter

Wiring diagram for the vehicle

NOTE: *An ohmmeter works by sending a small amount of current through the path to be measured. Because of this, all circuits and components being tested must be disconnected from power. An ohmmeter must never be connected to an energized circuit; doing so may damage the meter. The safest way to measure ohms is to disconnect the negative battery cable before taking resistance readings.*

Protective Clothing

Goggles or safety glasses with side shields

Describe the vehicle being worked on:

Year __________ Make ____________________ Model ____________________

VIN ____________________ Engine type and size ____________________

PROCEDURE

1. Locate the fuse panel or power distribution box.
2. With no power to the fuses, check the resistance of each fuse. Summarize your findings:

____

____

____

3. Now connect the leads of the analog meter across the negative battery cable.

   Your reading is: ________ ohms

   Now connect the leads of the digital meter across the negative battery cable.

   Your reading is: ________ ohms

4. Disconnect the wires leading to the ignition coil.

   Now connect the leads of the analog meter across the terminals of the coil.

   Your reading is: ________ ohms

   Now connect the leads of the digital meter across the terminals of the coil.

   Your reading is: ________ ohms

5. Reconnect the wires to the coil.

   Carefully remove one spark plug wire from the spark plug and the ignition coil or distributor cap.

   Now connect the leads of the analog meter across the wire.

   Your reading is: ________ ohms

   Now connect the leads of the digital meter across the wire.

   Your reading is: ________ ohms

6. Carefully reinstall the spark plug wire.

   Locate the cigar lighter inside the vehicle.

   Now connect the leads of the analog meter from the heating coil to its case.

   Your reading is: ________ ohms

   Now connect the leads of the digital meter from the heating coil to its case.

   Your reading is: ________ ohms

7. Refer to the service manual and find out how to remove the bulb in the dome light.

   Remove it.

   Now connect the leads of the analog meter across the bulb.

   Your reading is: ________ ohms

   Now connect the leads of the digital meter across the bulb.

   Your reading is: ________ ohms

8. Reinstall the bulb.

   Remove the rear brake light bulb.

   Now connect the leads of the analog meter across the bulb.

   Your reading is: ________ ohms

   Now connect the leads of the digital meter across the bulb.

   Your reading is: ________ ohms

9. Reinstall the bulb

   Disconnect the wire connector to one of the headlights.

   From the wiring diagram, identify which terminals are for low beam operation.

   Now connect the leads of the analog meter across the low beam terminals.

   Your reading is: ________ ohms

   Now connect the leads of the digital meter across the low beam terminals.

   Your reading is: ________ ohms

10. From the wiring diagram, identify which terminals are for high beam operation.

    Now connect the leads of the analog meter across the high beam terminals.

    Your reading is: ________ ohms

    Now connect the leads of the digital meter across the high beam terminals.

    Your reading is: ________ ohms

11. You should have gotten slightly different readings from the digital meter and the analog meter. Explain why:

    ____

    ____

    ____

    ____

12. You measured the resistance across several different light bulbs. On each you should have read a different amount of resistance. Based on your findings, which light bulb would be the brightest and which would be the dimmest? Explain why.

    ____

    ____

    ____

    ____

Problems Encountered

Instructor's Comments

ELECTRICAL AND ELECTRONIC SYSTEMS JOB SHEET 11

Testing Switches, Solenoids, Connectors, Relays and Wires

Name ______________________ Station ______________ Date ______________

NATEF Correlation:

This Job Sheet addresses the following NATEF tasks:

A.12. Check electrical circuits using fused jumper wires; determine necessary action.

A.16. Inspect and test switches, connectors, relays, solenoid solid state devices, and wires of electrical/electronic circuits; perform necessary action.

Objective

Upon completion of this job sheet, you will be able to inspect and test switches, connectors, relays, solenoids, and wires of electrical and electronic circuits with a DMM and jumper wires.

Tools and Materials

Test light

Fused jumper wire

Digital multimeter (DMM)

Scan tool

Protective Clothing

Goggles or safety glasses with side shields

Describe the vehicle being worked on:

Year ________ Make ____________________ Model ____________________

VIN ____________________ Engine type and size ____________________

PROCEDURE

Protection Devices

1. Check a fuse with an ohmmeter. If the fuse is good, there will be continuity through it. To test a circuit protection device with a voltmeter, check for available voltage at both terminals of the unit. If the device is good, voltage will be present on both sides. A test light can be used in place of a voltmeter. Summarize the results of this check.

____

____

2. Measure the voltage drop across a fuse or other circuit protection device. If a fuse, a fuse link, or circuit breaker is in good condition, a voltage drop of zero will be measured. If 12 volts is read, the fuse is open. Any reading between zero and 12 volts indicates some voltage drop. If there is a voltage drop across the fuse, it has resistance and should be replaced. Make sure you check the fuse holder for resistance as well. Summarize the results of this check.

____

____

Switches

1. To check a switch, disconnect the connector at the switch. With an ohmmeter, check for continuity between the terminals of the switch with the switch moved to the ON position and to the OFF position. With the switch in the OFF position, there should be no continuity between the terminals. With the switch on, there should be good continuity between the terminals. Summarize the results of this check.

____

____

2. If the switch is activated by something mechanical and doesn't complete the circuit when it should, check the adjustment of the switch. If the adjustment is correct, replace the switch. Task Completed ☐

3. Another way to check a switch is to simply bypass it with a jumper wire. If the component works when the switch is jumped, a bad switch is indicated. Summarize the results of this check.

____

____

4. Voltage drop across switches should also be checked. Ideally, when the switch is closed there should be no voltage drop. Any voltage drop indicates resistance and the switch should be replaced. Summarize the results of this check.

____

____

5. If the switch has multiple throws and/or poles, refer to the wiring diagram to identify what terminals should be connected in each switch position. Check for continuity across the terminals. There should be low resistance across only those terminals that should be connected while the switch is in that position. Make sure to do this in all switch positions. Summarize the results of this check.

____

____

Relays

1. Check the wiring diagram for the relay being tested to determine if the control is through an insulated or ground switch. If the relay is controlled on the ground side, continue with this procedure. If the relay is controlled on the positive side, describe the correct way to test it.

____

____

2. Connect the negative lead of the voltmeter to a good ground. Task Completed ☐

3. Connect the positive lead to the output wire. If no voltage is present, continue testing. If there is voltage, disconnect the ground circuit of the relay. The voltmeter should now read 0 volts. If it does, the relay is good. If voltage is still present, the relay is faulty and should be replaced. Task Completed ☐

4. Connect the positive voltmeter lead to the power input terminal. Close to battery voltage should be there. If not, the relay is faulty. If the correct voltage is present, continue testing. Task Completed ☐

5. Connect the positive meter lead to the control terminal. Close to battery voltage should be there. If not, check the circuit from the battery to the relay. If the correct voltage is there, continue testing. Task Completed ☐

6. Connect the positive meter lead to the relay ground terminal. If more than 1 volt is present, the circuit has a poor ground. Summarize the results of this check.

____

____

Solenoids

1. Disconnect the connector to the solenoid. Task Completed ☐

2. Measure resistance between the power terminal to the solenoid and body ground. What was your measurement and what does this indicate?

____

____

3. Connect a scan tool to the DLC. Task Completed ☐

4. Determine how to activate the solenoid being tested. Briefly explain the procedure.

____

____

5. Activate the solenoid and listen, did you hear a click? If so, what does this indicate? If not, what does this indicate?

____

____

Stepped Resistors

1. Remove the resistor from its mounting. Task Completed ☐
2. Make sure the ohmmeter is set to the correct scale for the anticipated amount of resistance. Connect the ohmmeter leads to the two ends of the resistor. Task Completed ☐
3. Compare the results against specifications. Summarize the results of this check.

____

____

Variable Resistors

1. Identify the input and output terminals and connect the ohmmeter across them. Task Completed ☐
2. Rotate or move the variable control while observing the meter. The resistance value should remain within the limits specified for the switch. If the resistance values do not match the specified amounts, or if there is a sudden change in resistance as the control is moved, the unit is faulty. Summarize the results of this check.

____

____

Problems Encountered

____

____

____

Instructor's Comments

____

____

____

ELECTRICAL AND ELECTRONIC SYSTEMS JOB SHEET 12

Test for Circuit Defects

Name ______________________ Station ______________ Date ______________

NATEF Correlation

This Job Sheet addresses the following NATEF task:

A.13. Locate shorts, grounds, opens, and resistance problems in electrical/electronic circuits; determine necessary action.

Objective

Upon completion of this job sheet, you will be able to locate shorts, shorts-to-ground, opens, and high resistance problems in electrical and electronic circuits.

Tools and Materials

Jumper wire
Digital multimeter (DMM)
Buzzer with fuse terminals
Cycling circuit breaker
Test light

Protective Clothing

Goggles or safety glasses with side shields

Describe the vehicle being worked on:

Year __________ Make ____________________ Model ____________________

VIN ____________________ Engine type and size ____________________

PROCEDURE

Testing for Opens

1. An open is usually indicated by an inoperative component or circuit. The easiest way to test a circuit is to start at the most accessible place and work from there. Is the load in the bad circuit accessible?

   ____

2. Check for voltage at the input or positive side of the load. If the reading was 10.5 volts or higher, check the ground side of the load. If the voltage there is 1 volt or lower and the load did not work, the load is bad. If the voltage at ground is greater than 1 volt, there is excessive resistance or an open in the ground circuit. If the voltage at the positive side of the load was less than 10.5 volts, proceed to the next step. But before you do, summarize your test results.

   ____

   ____

3. Move the positive lead of the voltmeter toward the battery, testing all connections along the way. If 10.5 volts or more are present at any connector, there is an open between that point and the point previously checked. Use a jumper wire to verify the location of the open. Summarize your test results.

____

____

4. If battery voltage was present at the ground of the load, there is an open in the ground circuit. Use a jumper wire to verify this. Summarize your test results.

____

____

Testing for Shorts

1. Use an ohmmeter to check for an internal short in a component. If the component is good, the meter will read the specified resistance or at least some resistance. If it is shorted, it will read lower than normal or zero resistance. Summarize your test results.

____

____

____

2. If the short is between circuits, check the wiring of the affected circuits for signs of burned insulation and melted conductors. Also check common connectors that are shared by the two affected circuits. Summarize your test results.

____

____

____

3. If a visual inspection does not identify the cause of a wire-to-wire short, remove one of the fuses for the affected circuits. Install a buzzer that has been fitted with terminals across the fuse holder terminals. Activate the circuit that the buzzer is connected to. Disconnect the loads that are supposed to be activated by the switch. Then disconnect the wire connectors in the circuit that connects the load to the switch. If the buzzer stops when a connector is disconnected, the short is in that circuit. Summarize your test results.

____

____

4. If the problem is a short to ground, the circuit's fuse or other protection device will be open. If the circuit is not protected, the wire, connector, or component will be burned or melted. As an aid to keep current flowing in the circuit so you can test it, connect a cycling circuit breaker across the fuse holder. The circuit breaker will continue to cycle open and closed, allowing you to test for voltage in the circuit. Task Completed ☐

5. Connect a test light in series with the cycling circuit breaker. While observing the test light, disconnect individual circuits and components one at a time until the light stays out. The short is in the circuit that was disconnected when the light went out. Summarize your test results.

____

____

____

Testing for High Resistance

1. High resistance problems are typically caused by corrosion on terminal ends, loose or poor connections, or frayed and damaged wires. Carefully inspect the affected circuit for these flaws. If the voltage drop is excessive, that part of the circuit contains the resistance. If the voltage drop was normal, the high resistance is in the switch or in the circuit feeding the switch. Summarize your test results.

____

____

2. Whenever excessive resistance is suspected, both sides of the circuit should be checked. Begin by checking the voltage at the positive side of the load. This should be close to battery voltage unless the circuit contains a resistor located before the load. If the voltage is less than desired, check the voltage drop across the circuit from the switch to the load. If the voltage drop is excessive, that part of the circuit contains the resistance. If the voltage drop was normal, the high resistance is in the switch or in the circuit feeding the switch. Summarize your test results.

____

____

3. Check the voltage drop across the switch. If the voltage drop was excessive, the problem is the switch. If the voltage drop was normal, the high resistance is in the circuit feeding the switch. Summarize your test results.

____

____

4. If battery voltage was present at the load, the ground circuit for the load should be checked. Connect the red voltmeter lead to the ground side of the load and the black lead to the grounding point for the circuit. If the voltage drop was normal, the problem is the grounding point. If the voltage drop was excessive, move the black meter lead toward the red. Check the voltage drop at each step. Eventually you will read high voltage drop at one connector and then low drop at the next. The point of high resistance is between those two test points. If the voltage drop was excessive, that part of the circuit contains the resistance. If the voltage drop was normal, the high resistance is in the switch or in the circuit feeding the switch. Summarize your test results.

____

____

____

Problems Encountered

____

____

____

Instructor's Comments

____

____

____

ELECTRICAL AND ELECTRONIC SYSTEMS JOB SHEET 13

Inspect and Check Fuses, Fusible Links, and Circuit Breakers

Name ______________________________ Station _________________ Date _________________

NATEF Correlation

This Job Sheet addresses the following NATEF task:

A.15. Inspect and test fusible links, circuit breakers, and fuses; determine necessary action.

Objective

Upon completion of this job sheet, you will be able to inspect and test fusible links, circuit breakers, and fuses.

Tools and Materials

Digital Multimeter (DMM)
Test light
Service manual

Protective Clothing

Goggles or safety glasses with side shields

Describe the vehicle being worked on:

Year __________ Make ________________________ Model ________________________

VIN ________________________ Engine type and size ________________________

PROCEDURE

1. Visually check fuses. Task Completed ☐
2. Using a test light to test a fuse, ground the clip on one end of the test light. Touch the probe to each end of the fuse. If the light comes on at both ends of the fuse, it is good. If the light works only when touched to one end of the fuse, it is bad. Replace it. Task Completed ☐
3. Using a test light to test a fusible link, ground the clip on one end of the test light. Touch the probe to each end of the fusible link. If the light comes on at both ends of the fusible link, it is good. If the light only comes on at one end of the fusible link, it is bad. Replace it. Task Completed ☐
4. Using a DMM to test a fuse or fusible link can be done two different ways. Task Completed ☐
5. If you are using the voltage scale, you will connect the black lead (the negative lead) to a ground. Task Completed ☐
6. Touch the red lead, positive lead, to each end of the fuse or fusible link. Task Completed ☐

7. If the voltage reading is the same at both ends, the fuse or fusible link is good.
8. If you only get a voltage reading at one end, the fuse or fusible link is bad. Replace it. Task Completed ☐
9. If you are using the ohmmeter to test a fuse or circuit breaker, the procedure is different.
10. Remove the fuse or circuit breaker. Task Completed ☐
11. Touch the leads from the meter to each end of the fuse or circuit breaker. Task Completed ☐
12. If there is continuity, the fuse or circuit breaker is good. Task Completed ☐
13. If there is no continuity, the fuse or circuit breaker is bad. Replace it. Task Completed ☐

Problems Encountered

____

____

____

Instructor's Comments

____

____

____

ELECTRICAL AND ELECTRONIC SYSTEMS JOB SHEET 14

Connector and Wire Repairs

Name ______________________ Station ______________ Date ______________

NATEF Correlation:

This Job Sheet addresses the following NATEF tasks:

A.17. Remove and replace terminal end from connector.

A.18. Repair connectors and terminal ends.

A.19. Repair wiring harness (including LAN/CAN/BUS systems).

Objective

Upon completion of this job sheet, you will be able to repair wiring harnesses, terminal ends, and connectors.

Tools and Materials

Terminal repair kit
Crimping tool
Dielectric grease
Heat shrink tubing
Heat gun
Electrical tape

Protective Clothing

Goggles or safety glasses with side shields

Describe the vehicle being worked on:

Year ________ Make ____________________ Model ____________________

VIN ____________________ Engine type and size ____________________

PROCEDURE

NOTE: *The preferred way to connect wires or to install a connector is by soldering. However, using solderless connectors is also an accepted practice and is the focus of this job sheet.*

1. Check all connectors and terminals inside the connector for corrosion, dirt, and looseness. Describe your findings.

____

____

2. Check the wiring harness for frayed, broken, or oil soaked wires, and faulty insulation. Describe your findings.

____

____

CAUTION: *On some models, the SRS wires are in a separate harness. If the SRS harness is damaged, replace the harness, do not repair it. On other models, wire harnesses include yellow SRS wires. If any SRS wire is damaged, replace the entire harness; do not repair it.*

Removing a Terminal from a Connector

1. If a terminal needs to be replaced, it must first be removed from the connector. Carefully separate the two halves of the connector. Nearly all connectors have pushdown release type locks. Make sure these are not damaged when disconnecting the connectors. Task Completed ☐
2. Refer to the service manual for the correct way to release a terminal from the connector. Most connectors have a secondary lock that serves as a backup for the primary lock on the terminal. Make sure the procedure for releasing both locks is followed. Summarize that procedure.

   ____

   ____

   ____

Installing New Terminals

1. Carefully match the old terminal with a new one. Choose the correct replacement terminal based on the wire size range the terminal will accommodate. Task Completed ☐

   WARNING: *Always follow the vehicle manufacturer's wiring and terminal repair procedure given in the service manual. On some components and circuits, manufacturers recommend component replacement rather than wiring repairs.*

2. Depending on the size of the wire you are repairing, use the proper size slot in the crimping tool. Task Completed ☐
3. Strip the insulation off the end of the wire so the wire fits in the new terminal. If the wire has a wire seal, replace it with a new one. While stripping the wire, make sure not to cut any wire strands. Task Completed ☐
4. Position the terminal in the crimping tool slot with the solid portion of the terminal toward the anvil. Task Completed ☐
5. Insert the wire into the terminal. Sometimes it is necessary to twist the wire strands in order to insert the wire into the terminal. Task Completed ☐
6. Squeeze the tool with both hands until the stops make contact. Task Completed ☐
7. Inspect the quality of the wire crimp. If it has a poor crimp, cut it off and start over. Task Completed ☐
8. Insert the terminal into the connector. Make sure the wire seals are pushed all the way into the connector. Then, pull gently on the wires to make sure the terminal is locked into place. Task Completed ☐
9. Close or insert the secondary lock, if applicable, and reconnect the connector. Task Completed ☐

Installing Pigtail Terminals

1. Pigtail terminals (short pieces of wire with a factory-crimped terminal) are used when the wire is too short or when access to the connector is too restricted to make a terminal repair. Remove the damaged or faulty terminal from the connector. Task Completed ☐
2. Cut off the wire about an inch back from where it connects to the damaged or faulty terminal, then strip about half of the insulation off that piece. Task Completed ☐
3. Select a pigtail terminal that matches the original wire at both ends (same kind of terminal and same diameter wire). Task Completed ☐
4. Select the smallest splice connector that will fit onto the stripped end of the original wire. Task Completed ☐
5. Insert the pigtail terminal into the connector; push it in until it locks in place. Task Completed ☐
6. Lay the pigtail and the original wire side-by-side, and cut off both ends at once. When making more than one splice, do not cut each pigtail at the same location; the resulting "lump" of splice connectors would interfere with rewrapping the harness. Instead, cut the first pigtail close enough to the terminal to leave room to make each remaining cut about 3/4 inch (20 mm) farther down on the next pigtail. Task Completed ☐
7. Put the splice connector in the proper size slot in the crimper tool, slide it to one end (where the flare begins), and close the crimper handles far enough to hold it in place. Task Completed ☐
8. Insert one of the bare wires into the splice connector end that is in the crimper. Push the wire all the way into the splice connector, and squeeze the crimper handles until the jaws touch. Task Completed ☐
9. Crimp the other wire in the same way into the other end of the splice connector. Task Completed ☐
10. After crimping, pull gently on the wires to make sure they are secure in the connector. Were you successful? If not, what should you do?

    ____

    ____

11. Separate the other wires in the harness from the repaired wire(s), and shield them with nonflammable material. What did you use to shield the other wires?

    ____

    ____

12. Apply heat from the heat gun at the middle of the splice connector, and move the gun toward the ends as the tube shrinks. Apply heat evenly by applying the heat around the splice connector. Shrinking is complete when a small amount of sealant appears at each end of the tube. Task Completed ☐

Repairing Wires

NOTE: *It may be necessary to bypass a length of wire that is not accessible. Cut the wire before it enters the inaccessible portion and at the other end where it leaves the area. Install a replacement wire and reroute it to the load. Be sure to protect the wire by using straps, hangers, or grommets as needed.*

WARNING: *Never reroute wires when making repairs. Rerouting wires can result in induced voltages in nearby components. These stray voltages can interfere with the function of electronic circuits.*

1. Cut out the section of wire that needs to be replaced. Task Completed ☐
2. Cut a length of wire of the same size from a spool of wire. Cut the wire a few inches longer than the length cut out of the wiring harness. What size wire are you repairing? ______________________ Task Completed ☐
3. Select the appropriate size and type of terminal connector. Be sure it is suitable for the unit's connecting post or prongs and it has enough current carrying capacity for the circuit. Also, make sure it is heavy enough to endure normal wire flexing and vibration. Task Completed ☐
4. Use the correct stripping opening on the crimping tool for the gauge of the wire and remove enough wire insulation to allow the wire to completely penetrate the connector. Strip both ends of the new wire and the ends of the two wires (the circuit that being repaired) remaining in the wiring harness. Task Completed ☐
5. Place the replacement wire into the connector and crimp the connector. To get a proper crimp, place the open area of the connector facing toward the anvil. Make sure the wire is compressed under the crimp. Install a connector at both ends of the replacement wire. Task Completed ☐
6. If you are going to use heat shrink tubing, place the correct size of tubing over the ends of the replacement wire. Make sure the tubing is longer than the connector. Task Completed ☐
7. Insert the stripped ends of the wire in the harness into the connectors of the replacement wire and crimp them. Task Completed ☐
8. Use electrical tape or heat shrink tubing to seal the connection. If you are going wrap the connection with tape, clean the area that will be taped with isopropyl alcohol. This will allow for good adhesion. Also, make sure the tape is tightly wrapped over the connections. Did you seal the connections with heat shrink tubing or tape?

   ____

   ____

Splicing Twisted or Shielded Wiring

1. Twisted and shielded wires are used in computer circuits to protect those circuits from electrical noise. To splice this type of wire: locate and cut out the damaged section of wire. Task Completed ☐

2. Do not cut into the Mylar tape or the drain wire. Remove about 1 inch (25mm) of the outer jacket from the ends of the cable. Task Completed ☐
3. Unwrap the Mylar tape but do not remove it from the cable. Task Completed ☐
4. Untwist the wires and remove the insulation from the ends. Task Completed ☐
5. Use a splice clip to connect the wires and solder the splice. Task Completed ☐
6. Wrap the wires with the Mylar tape. Do not wrap the drain wire in the tape. Task Completed ☐
7. If the drain wire is cut, splice the wire and solder the connection. Task Completed ☐
8. Wrap the drain wire around the other wires and the Mylar tape. Task Completed ☐
9. Use electrical tape or heat shrink tubing to insulate the cable. Task Completed ☐

Problems Encountered

____

____

____

Instructor's Comments

____

____

____

ELECTRICAL AND ELECTRONIC SYSTEMS JOB SHEET 15

Soldering Two Copper Wires Together

Name ______________________ Station ______________ Date ______________

NATEF Correlation

This Job Sheet addresses the following NATEF task:

A.20. Perform solder repair of electrical wiring.

Objective

Upon completion of this job sheet, you will be able to make wire repairs using solder.

Tools and Materials

100-watt soldering iron
60/40 Rosin core solder
Crimping tool
Splice clip
Heat shrink tube
Heating gun

Protective Clothing

Goggles or safety glasses with side shields

PROCEDURE

1. Disconnect the fuse that powers the circuit being repaired. NOTE: If a fuse does not protect the circuit, disconnect the ground lead of the battery. What did you need to do?

   ____

   ____

2. Locate and cut out the damaged wire. Where is the wire and what circuit does it belong to?

   ____

   ____

3. Using the correct size stripper, remove about 1/2-inch of the insulation from both wires. Task Completed ☐

4. Now remove about 1/2-inch of the insulation from both ends of the replacement wire. The length of the replacement wire should be slightly longer than the length of the wire removed. What size wire are you inserting to make the repair?

   ____

   ____

5. Select the proper size splice clip to hold the splice. Task Completed ☐
6. Place the correct size and length of heat shrink tube over the two ends of the wire. Task Completed ☐
7. Overlap the two spliced ends and center the splice clip around the wires, making sure the wires extend beyond the splice clip in both directions. Task Completed ☐
8. Crimp the splice clip firmly in place. Task Completed ☐
9. Heat the splice clip with the soldering iron while applying solder to the opening of the clip. Do not apply solder to the iron. The iron should be 180 degrees away from the opening of the clip. Task Completed ☐
10. After the solder cools, check the integrity of the joint. How did you do that?

    ____

    ____
11. Slide the heat shrink tube over the splice. Task Completed ☐
12. Heat the tube with the hot air gun until it shrinks around the splice. Do not overheat the heat shrink tube. Describe any problems you faced:

    ____

    ____

    ____

Problems Encountered

____

____

____

Instructor's Comments

____

____

____

ELECTRICAL AND ELECTRONIC SYSTEMS JOB SHEET 16

Disconnecting the High-Voltage Circuit on a Hybrid Vehicle

Name ______________________ Station ______________ Date ______________

NATEF Correlation:

This Job Sheet addresses the following NATEF tasks:

A.21. Identify the location of hybrid vehicle high voltage circuit disconnect (service plug) and safety procedures.

B.8. Identify high voltage circuits of electric or hybrid electric vehicle and related safety precautions.

Objective

Upon completion of this job sheet, you will be able to locate and safely disconnect the high voltage circuit on a hybrid vehicle.

Tools and Materials

Service manual

Hybrid vehicle

Vinyl tape for insulation

Protective Clothing

Goggles or safety glasses with side shields

Insulated gloves that are dry and are not cracked, ruptured, torn, or damaged in any way.

Describe the vehicle being worked on:

Year ________ Make ____________________ Model ____________________

VIN ____________________ Engine type and size ____________________

Describe the general condition of the vehicle:

____

____

PROCEDURE

WARNING: *Unprotected contact with any electrically charged ("hot" or "live") high-voltage component could cause serious injury or death.*

1. Describe how this vehicle is identified as a hybrid.

____

____

2. After referring to the vehicle's owner's manual, briefly describe its operation.

    ____

    ____

    ____

    ____

3. How many volts are the batteries rated at?

    ____

4. Where are the batteries located?

    ____

    ____

5. How are the high-voltage cables labeled and identified?

    ____

    ____

6. What is used to provide short circuit protection in the high-voltage battery pack?

    ____

    ____

7. What isolates the high-voltage system from the rest of the vehicle when the vehicle is shut off?

    ____

    ____

8. Using the vehicle's service and/or owner's manual as a guide, list at least five precautions that must be adhered to when servicing this hybrid vehicle.

    ____

    ____

    ____

    ____

    ____

9. If the vehicle has been in an accident and some electrolyte from the batteries has leaked out, how should you clean up the spill?

    ____

    ____

    ____

    ____

10. Look under the hood and describe the components that are visible.

____

____

____

____

11. Without touching the high-voltage cables or components, describe the routing of the cables. Include what they appear to be connected from and to.

____

____

____

____

WARNING: *Never assume that a hybrid vehicle is shut off simply because it is silent. Make sure the ignition key is in your pocket and not in the ignition switch.*

12. Describe the procedure for totally isolating the high-voltage system from the vehicle. This often involves the removal of a service plug. Be sure to include the location of this plug in your description of the procedure.

____

____

____

____

____

____

____

____

____

____

13. Make two signs saying, "WORKING ON HIGH VOLTAGE PARTS. DO NOT TOUCH!" Attach one to the steering wheel, and set the other one near the parts you are working on. Task Completed ☐

14. After the service plug has been removed, how long should you wait before working around or on the high-voltage system?

____

____

____

Problems Encountered

____

____

____

Instructor's Comments

____

____

____

ELECTRICAL AND ELECTRONIC SYSTEMS JOB SHEET 17

Inspection of a Battery for Condition, State of Charge, and Capacity

Name ______________________ Station ______________ Date ______________

NATEF Correlation:

This Job Sheet addresses the following NATEF tasks:

B.1. Perform battery state-of-charge (conductance) test; determine needed service.

B.2. Perform battery capacity test; determine needed service.

Objective

Upon completion of this job sheet, you will be able to perform battery state-of-charge tests and battery capacity tests.

Tools and Materials

Snap-On MicroVat Tester or other conductance-type tester
Starting/Charging System tester (VAT-40 or similar)
Digital multimeter (DMM)
Service manual

Protective Clothing

Goggles or safety glasses with side shields

Describe the vehicle being worked on:

Year ________ Make ______________________ Model ______________________

VIN ______________________ Engine type and size ______________________

PROCEDURE

1. Describe the general appearance of the battery.

____

____

2. Describe the general appearance of the cables and terminals.

____

____

3. Check the tightness of the cables at both ends. Describe their condition.

____

____

4. Connect the positive lead of the DMM (set on DC volts) to the positive terminal of the battery. Task Completed ☐

5. Put the negative of the meter on the battery case and move it all around the top and sides of the case. What readings did you get on the voltmeter?

____

____

6. What is indicated by those readings?

____

____

7. Measure the voltage of the battery. Your reading was _______ volts.

8. What do you know about the condition and state of charge of the battery based on the visual inspection and the above tests?

____

____

9. Locate the CCA rating of the battery. What is the rating? ______

10. If using a typical Starting/Charging System tester with an internal load, connect it to the battery. Task Completed ☐

11. Based on the CCA, how much load should be put on the battery during the capacity test? _______ amps

12. Conduct the battery load test. The battery's voltage decreased to _____ volts after _____ seconds.

13. If you are using a conductance-type tester, connect it and set it to conduct a battery test. Task Completed ☐

14. Enter the CCA rating of the battery being tested. Start the test. Task Completed ☐

15. Describe the results of either battery load (capacity) test. Include in the results your service recommendations and the reasons for them:

____

____

Problems Encountered

____

____

____

Instructor's Comments

____

____

____

ELECTRICAL AND ELECTRONIC SYSTEMS JOB SHEET 18

Maintaining Electronic Memory

Name ______________________ Station ______________ Date ______________

NATEF Correlation

This Job Sheet addresses the following NATEF task:

B.3. Maintain or restore electronic memory functions.

Objective

Upon completion of this job sheet, you will be able to maintain and restore electronic memory functions.

Tools and Materials

Memory saver

Protective Clothing

Goggles or safety glasses with side shields

Describe the vehicle being worked on:

Year ________ Make ____________________ Model ____________________

VIN ____________________ Engine type and size ____________________

PROCEDURE

1. Before disconnecting the battery on a vehicle, record all memory settings that the driver has control over, such as the radio and clock. Describe the accessories the vehicle has that have a memory function.

   ____

   ____

2. Insert the memory saver tool into the cigar lighter outlet or voltage receptacle. Task Completed ☐
3. Disconnect the negative cable of the battery. Then do whatever service work is required. Task Completed ☐
4. Reconnect the negative battery cable. Task Completed ☐
5. Remove the memory saver tool. Task Completed ☐
6. Check the settings on the accessories noted in step 1. Task Completed ☐
7. Refer to the service manual and follow the procedures outlined for resetting the parameters of the computer. This normally consists of a test drive under prescribed operating modes. Summarize the requirements for the test drive.

   ____

   ____

   ____

Problems Encountered

Instructor's Comments

ELECTRICAL AND ELECTRONIC SYSTEMS JOB SHEET 19

Remove, Clean, and Replace a Battery

Name ______________________ Station ______________ Date ______________

NATEF Correlation

This Job Sheet addresses the following NATEF tasks:

B.4. Inspect, clean, fill, and replace battery.

B.6. Inspect and clean battery cables, connectors, clamps, and hold-down; repair or replace as needed.

Objective

Upon completion of this job sheet, you will be able to properly inspect, clean, fill, and replace a battery as well as inspect and clean battery cables, connectors, clamps, and hold-down.

Tools and Materials

Baking soda
Battery clamp puller
Battery cleaning wire brush
Battery strap or carrier
Box wrench (1/2-in.) or cable-clamp pliers
Conventional wire brush and rags
Fender covers
Masking tape or bright felt marker
Petroleum jelly
Service manual

Protective Clothing

Rubber gloves
Rubber apron
Face shield, goggles, or safety glasses with side shields

Describe the vehicle being worked on:

Year ________ Make ____________________ Model ____________________

VIN ____________________ Engine type and size ____________________

PROCEDURE

1. Place a fender cover around the work area. Task Completed ☐

2. Consult the appropriate service manual for precautions about computer controls. Task Completed ☐

3. Remove the negative (–) battery terminal. For connectors tightened with nuts and bolts, loosen the nut with a box wrench or cable-clamp pliers. Using ordinary pliers or an open-end wrench can cause problems. Always grip the cable while loosening the nut. This will prevent unnecessary pressure on the terminal post that could break it or loosen its mounting in the battery. If the connector does not lift easily off the terminal when loosened, use a clamp puller to free it. Prying with a screwdriver or bar strains the terminal post. Task Completed ☐
4. Loosen the positive (+) battery cable and remove it from the battery with a battery terminal puller. If both battery cables are the same color, it is wise to mark the positive cable so that you will connect it to the correct terminal later. All that is needed to mark the cable is masking or similar kind of tape. Although differences in terminal size are designed to prevent reversing the polarity of the battery, marking the positive cable is an extra safeguard. Task Completed ☐
5. Loosen and remove the battery holddown straps, cover, and heat shield. Task Completed ☐
6. Lift the battery from the battery tray using a battery strap or carrier. Task Completed ☐
7. Clean and inspect the battery tray using a solution of baking soda and water. Task Completed ☐
8. Clean the battery cable terminals with baking soda solution and a battery terminal brush. Use the external portion to clean the post and the internal portion for the terminal ends. Task Completed ☐
9. Using a battery strap or carrier, install the replacement battery (new or recharged) in the battery tray. Task Completed ☐
10. Install the battery cover or holddown straps, and tighten their attaching nuts and bolts. Be certain the battery cannot move or bounce, but do not overtighten. Task Completed ☐
11. If so equipped, reinstall heat shield. Task Completed ☐
12. Reinstall, beginning with the positive cable, both terminal connectors. Do not overtighten because this could damage the post or connectors. Coat the connectors with petroleum jelly. Task Completed ☐
13. Test the installation by starting the engine. Task Completed ☐

Problems Encountered

____

____

____

Instructor's Comments

____

____

____

ELECTRICAL AND ELECTRONIC SYSTEMS JOB SHEET 20

Charge a Maintenance-Free Battery

Name ____________________ Station ______________ Date ______________

NATEF Correlation

This Job Sheet addresses the following NATEF task:

B.5. Perform slow/fast battery charge.

Objective

Upon completion of this job sheet, you will be able to properly charge a battery.

Tools and Materials

Baking soda
Battery charger and cables and adapters for side terminal batteries
Battery clamp puller
Battery cleaning wire brush
Battery strap or carrier
Box wrench (1/2-in.) or cable-clamp pliers
Conventional wire brush and rags
Fender covers
Masking tape or bright felt marker
Petroleum jelly
Service manual
Voltmeter

Protective Clothing

Gloves
Rubber apron
Safety goggles, face shield, or safety glasses with side shields

PROCEDURE

CAUTION: *Special care must be given to charging maintenance-free batteries. If any electrolyte is lost during the charging process, the life of the battery is shortened. The fast-rate charge for this type of battery should be limited to 35 amperes for 20 minutes.*

1. Place fender covers around the work area. Remove battery from vehicle. It is also possible to charge a battery in the vehicle. If performing an in-vehicle charge, consult the service manual for any precautions associated with computer controls. Task Completed ☐

2. Check that the charger is turned off. Connect the positive (+) cable from the charger to the positive (+) battery terminal. Connect the (–) cable from the charger to the negative (–) battery terminal. Be sure you have a good connection to prevent sparking. Task Completed ☐

3. If at no-load the battery reads below 12.2 volts, charge the battery according to the following rates: Task Completed ☐

Battery Capacity (Reserve Minutes)	Slow Charge
80 minutes or less	10 hours at 5 amperes, or 5 hours at 10 amperes
Above 80–125 minutes	15 hours at 5 amperes, or 7-1/2 hours at 10 amperes
Above 125–170 minutes	20 hours at 5 amperes, or 10 hours at 10 amperes
Above 170–250 minutes	30 hours at 5 amperes, or 15 hours at 10 amperes
Above 250 minutes	20 hours at 10 amperes

If the voltage of the battery at room temperature is 12.2 volts, charge the battery for half the time shown under "Slow Charge." If the voltage is 12.4 volts, charge the battery for one-fourth the "Slow Charge" time.

Turn the clock control on the charger to the desired charging time. Do not exceed the manufacturer's battery-charging limits, which generally appear on the battery.

4. Charge the battery to a voltage of at least 12.6 volts or until the green ball appears. Shaking or tipping the battery may be necessary to make the green ball appear. Never overcharge a battery. Task Completed ☐

WARNING: *Never smoke around a charging battery. The hydrogen gas produced is highly explosive.*

5. When the battery is fully charged, turn off the power switch and disconnect the two battery charger cables from the battery. Return the charger to the proper area. Task Completed ☐

Problems Encountered

____

____

____

Instructor's Comments

____

____

____

ELECTRICAL AND ELECTRONIC SYSTEMS JOB SHEET 21

Jump-Starting a Vehicle

Name ______________________ Station ______________ Date ______________

NATEF Correlation

This Job Sheet addresses the following NATEF task:

B.7. Start a vehicle using jumper cables and a battery or auxiliary power supply.

Objective

Upon completion of this job sheet, you will be able to start a vehicle using jumper cables and a battery according to manufacturer's recommendations.

Tools and Materials

Jumper cables

Protective Clothing

Goggles or safety glasses with side shields

Describe the vehicle being worked on:

Year ________ Make ____________________ Model ____________________

VIN ____________________ Engine type and size ____________________

PROCEDURE

1. Make sure the two vehicles are not touching each other. Task Completed ☐
2. Apply the parking brake for each vehicle and put the transmissions in neutral or park. Task Completed ☐
3. Turn off the ignition switch and the accessories on both vehicles. Task Completed ☐
4. Attach one end of the positive jumper cable to the disabled battery's positive terminal. Task Completed ☐
5. Connect the other end of the positive jumper cable to the booster battery's positive terminal. Task Completed ☐
6. Attach one end of the negative jumper cable to the booster battery's negative terminal. Task Completed ☐
7. Attach the other end of the negative jumper cable to an engine ground on the disabled vehicle. Task Completed ☐
8. Attempt to start the disabled vehicle. If the disabled vehicle does not readily start, start the jumper vehicle and run it at fast idle. Task Completed ☐
9. Once the disabled vehicle starts, disconnect the ground connected negative jumper cable from the engine block. Task Completed ☐

10. Disconnect the negative jumper cable from the booster battery. Task Completed ☐
11. Disconnect the positive jumper cable from the booster battery, then from the other battery. Task Completed ☐

Problems Encountered

____

____

____

Instructor's Comments

____

____

____

ELECTRICAL AND ELECTRONIC SYSTEMS JOB SHEET 22

Memory Resets

Name ______________________ Station ______________ Date ______________

NATEF Correlation:

This Job Sheet addresses the following NATEF task:

B.9. Identify electronic modules, security systems and/or radios that require reinitialization or code entry following battery disconnect.

Objective

Upon completion of this job sheet, you will be able to identify what systems will be affected by a memory loss after the vehicle's battery has been disconnected. You will also be able to prepare for resetting the memory before disconnecting a battery.

Tools and Materials

Service manual

Protective Clothing

Goggles or safety glasses with side shields

Describe the vehicle being worked on:

Year ________ Make ____________________ Model ____________________

VIN ____________________ Engine type and size ____________________

Describe the accessories the vehicle is equipped with:

____

____

____

____

PROCEDURE

1. Depress the memory button for the driver's power seat. This will put the seat and possibly outside mirrors in the desired position. While working on the vehicle, do not change the position of the seat. Task Completed ☐

2. Turn on the radio. List of all the set AM stations with their button number:

____

____

3. Now, list all of the FM radio stations and their button number (remember many radios have multiple FM ranges, sometimes referred to as FM1, FM2, etc.), make sure you record all set stations.

____

____

____

____

____

____

4. If the radio has an anti-theft feature, make sure you have the code. Where did you get the code? What is it?

____

____

____

5. Secure the code for the vehicle's anti-theft system. Check the owner's manual for the procedure for entering this code or for any special procedures that need to be followed when disconnecting the battery. Briefly describe the procedure.

____

____

6. On most vehicles, the battery can now be disconnected. Check the service manual to see if anything else should be recorded prior to disconnecting the battery. Record your findings here:

____

____

7. Once the battery is reconnected, enter the vehicle's anti-theft code or follow the procedure for resetting the alarm system. Task Completed ☐

8. Enter the radio anti-theft code. What is the procedure for doing this?

____

____

9. Now enter AM and FM stations previously recorded. What is the procedure for doing this?

____

____

10. Using an accurate source as a reference, set the time on the clock. What is the procedure for doing this?

____

____

11. Are there any other features of the radio or clock that need to be reset, such as refreshing a satellite radio?

____

____

12. If the power seat has not been moved since the before the battery was disconnected, set the memory for the seat. What is the procedure for doing this?

____

____

13. Using the service manual as a guide, determine if the memory of these other systems will need to be reset. Briefly describe the procedure for reinitializing the systems.

A. Engine controls

____

____

B. Idle learn

____

____

C. Automatic transmission

____

____

D. Suspension system

____

____

E. Tire pressure monitoring system

____

____

F. Power windows

____

____

G. Power hatch

____

____

H. Sliding roof

____

____

14. Are there any other precautions that must be followed before returning the vehicle to its owner?

____

____

Problems Encountered

____

____

____

Instructor's Comments

____

____

ELECTRICAL AND ELECTRONIC SYSTEMS JOB SHEET 23

Servicing the Auxiliary Battery in a Hybrid Vehicle

Name ______________________ Station ______________ Date ______________

NATEF Correlation:

This Job Sheet addresses the following NATEF task:

B.10. Identify hybrid vehicle auxiliary (12v) battery service, repair and test procedures.

Objective

Upon completion of this job sheet, you will be able to identify all special service procedures that relate to the servicing and testing of the auxiliary battery used in hybrid vehicles.

Tools and Materials

Service manual

Hybrid vehicle

Protective Clothing

Goggles or safety glasses with side shields

Describe the vehicle being worked on:

Year ________ Make ____________________ Model ____________________

VIN ____________________ Engine type and size ____________________

PROCEDURE

1. According to the service manual and/or owner's manual, how many batteries does this vehicle have? What are their voltage ratings?

   ____

   ____

2. Where are these batteries located?

   ____

   ____

3. What is the purpose of the auxiliary battery?

   ____

   ____

4. Are there any special service or test procedures required for this battery? If so, what are they?

____

____

____

Problems Encountered

____

____

____

Instructor's Comments

____

____

____

____

ELECTRICAL AND ELECTRONIC SYSTEMS JOB SHEET 24

Testing the Starter and Starter Circuit

Name ______________________ Station ______________ Date ______________

NATEF Correlation

This Job Sheet addresses the following NATEF tasks:

C.1. Perform starter current draw tests; determine necessary action.

C.2. Perform starter circuit voltage drop tests; determine necessary action.

C.3. Inspect and test starter relays and solenoids; determine necessary action.

Objective

Upon completion of this job sheet, you will be able to perform starter current draw tests and starter circuit voltage drop tests, as well as inspect and test starter relays and solenoids.

Tools and Materials

Fender covers
Remote starter switch
Service manual
Digital multimeter (DMM)
Starting/Charging System tester (VAT-40 or similar)

Protective Clothing

Goggles or safety glasses with side shields

Describe the type of vehicle being worked on:

Year ________ Make ____________________ Model ____________________

VIN ____________________ Engine type and size ____________________

PROCEDURE

1. Draw a simple starter circuit for the vehicle. Use the wiring diagram as a guide.

2. Place fender covers on the fenders of the vehicle. Task Completed ☐

3. Disable the ignition or fuel injection to prevent the engine from starting. Task Completed ☐

   Instructor's Check: ____________

4. If available, connect a remote starter switch. Task Completed ☐

5. Connect the voltmeter across the battery's negative cable. Task Completed ☐

   Crank the engine with the starter and observe the readings on the meter. Task Completed ☐

   Your reading was __________ volts.

6. What does this indicate?

   ____

   ____

7. Connect the voltmeter across the battery's positive cable (from battery to starter motor). Task Completed ☐

   Crank the engine with the starter and observe the readings on the meter. Your reading was __________ volts.

8. This test measured the voltage drop across everything in the positive side of the circuit. What is included in this circuit?

   ____

   ____

   ____

9. What do the test results suggest? Where should the meter be connected to next to identify the cause of any excessive voltage drop?

   ____

   ____

   ____

10. Refer to the service manual for normal starter current draw. Task Completed ☐

11. Expected starter current draw is __________ amps

    Voltage should not drop below __________ volts

12. Connect the Starting/Charging System tester cables to the vehicle. Task Completed ☐

13. Zero the ammeter on the tester. Task Completed ☐

14. Be prepared to observe the amperage when the engine begins to crank and while it is cranking. Also note the voltage when you stop cranking the engine. Task Completed ☐

    The initial current draw was __________ amps

    After __________ seconds, the current draw was __________ amps and the voltage dropped to __________ volts.

15. What is indicated by the test results? Compare your measurements to the specifications.

____

____

____

____

____

16. Reconnect the ignition or fuel injection system and start the engine. Task Completed ☐

Problems Encountered

____

____

____

Instructor's Comments

____

____

____

ELECTRICAL AND ELECTRONIC SYSTEMS JOB SHEET 25

Remove, Inspect, and Replace a Starter Motor

Name ______________________ Station ______________ Date ______________

NATEF Correlation

This Job Sheet addresses the following NATEF task:

C.4. Remove and install starter in a vehicle.

Objective

Upon completion of this job sheet, you will be able to properly remove and install a starter.

Tools and Materials

Appropriate screwdrivers
Appropriate wrenches
Cleaning solvent
Fender covers
Hoist or safety stands
Lubricant
Rag
Service manual
Support brackets (if applicable)
Tags

Protective Clothing

Goggles or safety glasses with side shields

Describe the vehicle being worked on:

Year ________ Make ____________________ Model ____________________

VIN ____________________ Engine type and size ____________________

PROCEDURE

1. Place fender covers on the vehicle. Disconnect the battery ground cable. Task Completed ☐

   WARNING: *Never attempt to remove the starter without first disconnecting the battery.*

2. Raise and support the vehicle using a hoist or safety stands. Task Completed ☐
3. If required to access the starter, remove the front wheels. Loosen or remove the exhaust pipes and other components that block access to the starter. Task Completed ☐
4. Disconnect and tag all wires from the starter motor and/or solenoid. Task Completed ☐
5. Remove the heat shields and support brackets. Remove all mounting bolts and shims that secure the starter to the engine, and lift out the starter. Task Completed ☐
6. Visually inspect the starter bushing and armature shaft bearing area. Inspect the pinion gear and overrunning clutch drive unit. Check the pinion housing for cracks. If a problem is found, replace the starter motor. Record your results on the Report Sheet for Starter Motor Inspection and Replacement. Task Completed ☐

7. Check the overrunning clutch pinion clearance by either forcing (where possible) the solenoid arm to the fully applied position or by energizing the starter solenoid. Task Completed ☐

8. When the solenoid has moved the pinion to the fully engaged position, push the pinion back to remove all slack. Measure the clearance between pinion and pinion stop. Record the results on the Report Sheet for Starter Motor Inspection and Replacement found at the end of this Job Sheet. Task Completed ☐

9. Lubricate the starter bushing and armature shaft with a few drops of 20-weight oil. Task Completed ☐

10. A rag, slightly dampened with clean solvent, may be used to wipe the starter housing if it is oily. If no oil is present, use a soft brush or a clean, dry rag. Task Completed ☐

11. Attach the heat shield and support brackets to the new or serviced starter motor. Some starters use a sealer in some locations to prevent entry of dust and water. When assembling parts, apply a nonhardening-type sealer. Where grommets are used, they must be in good condition and properly inserted. Task Completed ☐

12. Install the motor to the engine using all mounting bolts. Torque the bolts to specifications. Task Completed ☐

13. Check for proper flywheel engagement. Many starter designs have no adjustment for pinion clearance. When it is wrong, correct it by installing new parts. Task Completed ☐

14. Reconnect all wires to the solenoid and/or motor terminals. Remount or tighten any removed or loosened parts. Task Completed ☐

15. Lower the vehicle and reconnect the battery ground cable. Test the starter for proper operation. Task Completed ☐

16. Remove the fender covers. Clean work area, and return vehicle to designated area. Task Completed ☐

Problems Encountered

____

____

____

Instructor's Comments

____

____

____

Name ______________________ Station ______________ Date ______________

REPORT SHEET FOR STARTER MOTOR INSPECTION AND REPLACEMENT		
	Serviceable	*Not Serviceable*
1. Visual inspection		
Bushing		
Armature shaft bearing		
Pinion gear		
Overrunning clutch		
Pinion housing		
Flywheel ring gear teeth		
	Clearance/Torque	
2. Actual pinion clearance		
3. Starter-motor-to-block-bolt torque		

Conclusions and Recommendations ____

____

____

ELECTRICAL AND ELECTRONIC SYSTEMS JOB SHEET 26

Free Speed (No-Load) Starter Motor Test

Name ______________________ Station ______________ Date ______________

NATEF Correlation

This Job Sheet addresses the following NATEF task:

C.1. Perform starter current draw tests; determine necessary action.

Objective

Upon completion of this job sheet, you will be able to correctly perform a starter bench test.

Tools and Materials

Vise	Battery cables
Ammeter with inductive pickup	Battery
Digital multimeter (DMM)	Service manual

Protective Clothing

Goggles or safety glasses with side shields

Describe the vehicle being worked on:

Year ________ Make ____________________ Model ____________________

VIN ____________________ Engine type and size ____________________

PROCEDURE

1. Remove the starter from the vehicle. Task Completed ☐
2. Clamp the starter firmly in a bench vise. Task Completed ☐
3. What are the no-load specifications for this starter motor?

   ____

4. Connect the positive cable from the battery to the starter's battery connection. Task Completed ☐
5. Put the inductive lead of the ammeter over the negative cable. Task Completed ☐
6. Connect the negative cable from the battery to the frame of the starter. Task Completed ☐
7. Is the motor running? Describe its sound.

   ____

   ____

8. Check the battery's voltage and the current draw and speed of the motor and compare them to specifications. Summarize the results.

____

____

____

9. Based on the results of this test, what are your service recommendations?

____

____

____

Problems Encountered

____

____

____

Instructor's Comments

____

____

____

ELECTRICAL AND ELECTRONIC SYSTEMS JOB SHEET 27

Control Circuit Test

Name ______________________________ Station __________________ Date __________________

NATEF Correlation

This Job Sheet addresses the following NATEF task:

C.5. Inspect and test switches, connectors, and wires of starter control circuits; perform necessary action.

Objective

Upon completion of this job sheet, you will be able to inspect and test switches, connectors, and wires of starter control circuits.

Tools and Materials

Fender covers

Service manual

Voltmeter

Protective Clothing

Goggles or safety glasses with side shields

Describe the vehicle being worked on:

Year __________ Make _________________________ Model _________________________

VIN _________________________ Engine type and size _________________________

PROCEDURE

1. Place fender cover on fender. Check vehicle wiring diagram, if possible, to identify all control circuit components. These normally include ignition switch, safety switch, starter solenoid winding, and/or a separate relay. Task Completed ☐
2. Disable the ignition or fuel injection system so the engine will not start when the starter is activated. Task Completed ☐
3. Connect the red lead of the voltmeter to the positive terminal of the battery and the black lead to the ignition switch terminal at the solenoid or starter relay. Task Completed ☐
4. Crank the engine with the starter and observe the voltage reading. Your reading was _____ volts.
5. What do you conclude from this test?

____

____

6. Normally, if you were diagnosing a system and you measured less than 0.1 volts during the last test, you would stop there and move on to testing another part of the starting system. For the purposes of this job sheet, proceed through the rest of the procedure.
7. Connect the voltmeter across the following circuits and record the voltage reading while the engine is being cranked.
   a. Battery (or fuse block) to ignition switch: ________ volts
   b. Input to output of the ignition switch: ______ volts
   c. Output of ignition switch to safety switch input: ______ volts
   d. Input of safety switch to output: _____ volts
   e. Output of safety switch to solenoid or relay: _____ volts
8. A reading of more than 0.1 volt across any one wire or switch usually indicates trouble. If a high reading is obtained across a safety switch used on automatic transmissions, check the adjustment of the switch according to manufacturer's service manual. Task Completed ☐
9. Remove fender covers. Clean work area, and return vehicle to designated area. Task Completed ☐

Problems Encountered

____

____

____

Instructor's Comments

____

____

____

ELECTRICAL AND ELECTRONIC SYSTEMS JOB SHEET 28

Typical Procedure for Disassembling and Reassembling a Starter

Name ______________________ Station ______________ Date ______________

NATEF Correlation

This Job Sheet addresses the following NATEF task:

C.4. Remove and install starter in a vehicle.

Objective

Upon completion of this job sheet, you will be able to disassemble, clean, inspect, test, and reassemble a starter motor.

Tools and Materials

Rags	Plastic head hammer
Assorted wrenches	Punch
Snap-ring pliers	Scribe
Flat blade screwdriver	Small press
Ball peen hammer	

Protective Clothing

Goggles or safety glasses with side shields

Describe the vehicle being worked on:

Year ________ Make ____________________ Model ____________________

VIN ____________________ Engine type and size ____________________

PROCEDURE

1. With the starter motor removed, clean the case. Task Completed ☐
2. Scribe reference marks at each end of the starter end housings and the frame. Task Completed ☐
3. Disconnect the field coil connection at the solenoid. What terminal of the solenoid is this?

   ____
4. Remove the screws that attach the solenoid to the starter drive housing. Task Completed ☐
5. Rotate the solenoid until the locking flange of the solenoid is free. Then remove the solenoid. Task Completed ☐
6. Remove the through-bolts from the end frame. Task Completed ☐
7. Remove the end frame and frame of the starter. Task Completed ☐

8. Remove the armature. What did you need to do in order to get the armature out?

____

____

9. Place a deep socket over the armature shaft until it contacts the retaining ring of the starter drive. What size socket did you use?

____

10. Tap the end of the socket with a plastic-faced hammer to drive the retainer toward the armature. Move it only far enough to access the snap ring. Task Completed ☐

11. Remove the snap ring. Task Completed ☐

12. Remove the retainer from the shaft and remove the clutch and spring from the shaft. Press out the drive housing bushing and the end-frame bushing. Task Completed ☐

13. Inspect all parts and describe the condition of the following:

    Brushes: ____

    Armature: ____

    Commutator: ____

    Frame: ____

    Bearings: ____

    Starter drive: ____

    Field coils: ____

14. With an ohmmeter, check the armature for shorts, opens, and high resistance. How did you do that and what were your results?

____

____

____

____

____

15. With an ohmmeter, check the field windings for shorts, opens, and high resistance. How did you do that and what were your results?

____

____

____

____

16. Summarize your findings about the parts of the starter. What parts can be reused? What parts need to be replaced?

____

____

____

____

17. Gather all of the required new parts, then press in the drive housing bushing and the end-frame bushing, using the appropriate drivers. Task Completed ☐

18. Install the clutch and spring onto the armature shaft and install the retainer. Task Completed ☐

19. Place the deep socket over the armature shaft until it contacts the retaining ring of the starter drive. Tap the end of the socket with a plastic-faced hammer to drive the retainer toward the armature. Move it only far enough to install a new snap ring. Task Completed ☐

20. Install the armature into the frame. What did you need to do in order to get the armature in?

____

____

21. Install the end frame over the armature shaft. Make sure the brushes are in contact with the commutator. Also make sure the end frame is aligned with the alignment marks you made during disassembly. What did you need to do in order to get the commutator past the brushes?

____

____

____

22. Install and tighten the through-bolts from the end frame. Task Completed ☐

23. Rotate the solenoid until it is locked in place. Task Completed ☐

24. Install and tighten the screws that attach the solenoid to the starter drive housing. Task Completed ☐

25. Reconnect the field coil connection at the solenoid. Task Completed ☐

26. Bench test the starter. How did you do that and what were the results?

____

____

____

Problems Encountered

Instructor's Comments

ELECTRICAL AND ELECTRONIC SYSTEMS JOB SHEET 29

Slow/No-Start Diagnosis

Name ______________________ Station ______________ Date ______________

NATEF Correlation

This Job Sheet addresses the following NATEF task:

C.6. Differentiate between electrical and engine mechanical problems that cause a slow-crank or no-crank condition.

Objective

Upon completion of this job sheet, you will be able to determine if the cause of a slow-start or a no-start condition is the engine or something in the starter or its electrical circuit.

Tools and Materials

Assorted Wrenches

Tachometer

Protective Clothing

Goggles or safety glasses with side shields

Describe the vehicle being worked on:

Year ________ Make ____________________ Model ____________________

VIN ____________________ Engine type and size ____________________

PROCEDURE

NOTE: *Most of these questions can be answered whether or not the vehicle actually has a cranking problem. If it does not, don't answer the specific questions but be sure to answer the general questions.*

1. Prior to testing the starting system when there is a no-start or slow-start condition, certain checks should be made. The first of which is the state-of-charge of the battery. Measure the voltage of the battery and record your findings here.

 ____

 ____

2. Does this voltage rating indicate at least a three-quarter charged battery? If not, what should you do before continuing your diagnosis?

 ____

 ____

 ____

3. Connect a tachometer to the engine. Task Completed ☐
4. Crank the engine and observe the tachometer. What did you observe?

___

5. Was the engine speed during cranking between 200–400 rpm? What does that indicate?

___

___

6. If the speed was below 200 or if the engine didn't crank, what could be the problem?

___

___

___

___

7. If the engine speed was above 400 rpm, explain what is going on.

___

___

___

___

8. If the engine did not crank, turn the crankshaft pulley in a clockwise direction with the appropriate wrench and/or socket. Describe your results.

___

___

9. What is indicated by the results of your tests? If the engine didn't rotate or if it was very difficult to rotate it, what could be the cause?

___

___

10. If the engine had the problems you cited in answer #9, would that be enough to stop the starter motor from rotating the engine? Why?

___

___

Problems Encountered

Instructor's Comments

ELECTRICAL AND ELECTRONIC SYSTEMS JOB SHEET 30

Charging System Tests

Name ______________________ Station ______________ Date ______________

NATEF Correlation

This Job Sheet addresses the following NATEF tasks:

D.1. Perform charging system output test; determine necessary action.

D.2. Diagnose charging system for the cause of undercharge, no-charge, and overcharge conditions.

D.5. Perform charging circuit voltage drop tests; determine necessary action.

Objective

Upon completion of this job sheet, you will be able to perform a charging system output test, diagnose a charging system for the cause of undercharge, no-charge, and overcharge conditions as well as perform charging circuit voltage drop tests.

Tools and Materials

Service manual and wiring diagram for the vehicle

Digital multimeter (DMM)

Starting/Charging System tester (VAT-40 or similar)

Protective Clothing

Goggles or safety glasses with side shields

Describe the vehicle being worked on:

Year ________ Make ____________________ Model ____________________

VIN ____________________ Engine type and size ____________________

PROCEDURE

1. Describe the general appearance of the AC generator and the wires that are attached to it.

____

____

____

____

2. Measure the open circuit voltage of the battery. Your measurement was ______ volts.

3. From the wiring diagram, identify the output, input, and ground wires for the AC generator. Describe these wires, by color and location, below.

____

____

____

____

4. Start the engine and allow it to run. Then turn on the headlights in the high-beam mode.

5. Connect the DMM across the charging system's output wire. Measure the voltage drop and record your readings.

____

6. Connect the DMM across the charging system's input wire. Measure the voltage drop and record your readings.

____

7. Connect the DMM across the charging system's ground wire. Measure the voltage drop and record your readings.

____

8. Based on the above tests, what is indicated by the results?

____

____

____

____

9. Identify the type and model of AC generator. What type is it?

____

What are the output specifications for this AC generator?

________ amps and ________ volts at __________ rpm

10. Connect the Starting/Charging System tester to the vehicle.

11. Start the engine and run it at the specified engine speed.

12. Observe the output to the battery. The meter readings are:

_________ amps and _______ volts

13. Compare readings to specifications and give recommendations.

____

____

____

____

14. If readings are outside of the specifications, refer to the service manual to find the proper way to full field the AC generator. Describe the method.

____

____

____

____

15. Full field the generator and observe the output to the battery. The meter readings are: ___________ amps and ________ volts

16. Compare readings to specifications and give recommendations.

____

____

____

____

____

Problems Encountered

____

____

____

Instructor's Comments

____

____

____

ELECTRICAL AND ELECTRONIC SYSTEMS JOB SHEET 31

Inspect, Replace, and Adjust Drive Belts and Pulleys

Name ______________________ Station ______________ Date ______________

NATEF Correlation

This Job Sheet addresses the following NATEF task:

D.3. Inspect, adjust, or replace generator (alternator) drive belts, pulleys, and tensioners: check pulley and belt alignment.

Objective

Upon completion of this job sheet, you will be able to inspect and adjust a generator (alternator) drive belt.

Tools and Materials

Two vehicles, one with a serpentine belt and the other with V-belts

Service manuals for the above vehicles

Belt tension gauge

Hand tools

Protective Clothing

Goggles or safety glasses with side shields

Describe the vehicle being worked on:

Year ________ Make ____________________ Model ____________________

VIN ____________________ Engine type and size ____________________

PROCEDURE

1. On the vehicle with a serpentine belt, carefully inspect the belt and describe the general condition of the belt.

   ____

   ____

   ____

   ____

2. With the proper belt tension gauge, check the tension of the belt. Belt tension should be ________. You found ______________.

3. Based on the above, what are your recommendations?

____

____

4. Describe the procedure for adjusting the tension of the belt:

____

____

5. Referring to the service manual, describe the procedure for removing the belt:

____

____

6. Draw the proper routing for installing a new drive belt:

7. On the vehicle with V-belts, you will find more than one drive belt. List the different belts by their purpose.

____

____

____

____

8. Carefully inspect the belts and describe the general condition of each one.

____

____

____

____

9. Check the tension of the AC generator drive belt. Belt tension should be ________________. You found ________________.

10. Based on the above, what are your recommendations?

____

____

11. Describe the procedure for adjusting the tension of the belt:

____

____

____

____

12. Describe the procedure for removing an AC generator drive belt:

____

____

____

____

Problems Encountered

____

____

____

Instructor's Comments

____

____

____

ELECTRICAL AND ELECTRONIC SYSTEMS JOB SHEET 32

Checking the Voltage Regulator

Name ______________________ Station ______________ Date ______________

NATEF Correlation

This Job Sheet addresses the following NATEF task:

D.2. Diagnose charging system for the cause of undercharge, no-charge, and overcharge conditions.

Objective

Upon completion of this job sheet, you will be able to inspect and test a voltage regulator/regulation circuit.

Tools and Materials

Digital multimeter (DMM)

Tachometer

Protective Clothing

Goggles or safety glasses with side shields

Describe the vehicle being worked on:

Year ________ Make ____________________ Model ____________________

VIN ____________________ Engine type and size ____________________

PROCEDURE

1. Check the alignment, condition, and tension of the generator's drive belt. Describe your findings.

____

____

2. Inspect the battery and the cables and terminals connected to it. Describe your findings.

____

____

3. Inspect all charging system wiring and connectors. Describe your findings.

____

____

4. Inspect the AC generator and regulator mountings for loose or missing bolts. Task Completed ☐

5. Connect a voltmeter across the battery terminals. Task Completed ☐

6. Connect a tachometer if the vehicle doesn't have one. Task Completed ☐

7. With the engine off, measure and record the voltage at the battery. Describe your findings.

____

____

8. Start the engine and bring its speed to about 1500 to 2000 rpm. Task Completed ☐

9. Measure and record the voltage at the battery. It should now be 13.5 to 14.5 volts. Describe your findings.

____

____

10. If the charging voltage was too high, describe what may be the cause.

____

____

11. If the charging voltage was too low, describe what may be the cause.

____

____

12. Explain how this test checks the operation of the voltage regulator.

____

____

____

Problems Encountered

____

____

____

Instructor's Comments

____

____

____

ELECTRICAL AND ELECTRONIC SYSTEMS JOB SHEET 33

Remove and Replace an AC Generator

Name ______________________ Station ______________ Date ______________

NATEF Correlation

This Job Sheet addresses the following NATEF task:

D.4. Remove, inspect, and install generator (alternator).

Objective

Upon completion of this job sheet, you will be able to properly remove, inspect, and install a generator (alternator).

Tools and Materials

Appropriate wrenches
Screwdrivers
Fender cover
Service manual

Protective Clothing

Goggles or safety glasses with side shields

Describe the vehicle being worked on:

Year ________ Make ____________________ Model ____________________

VIN ____________________ Engine type and size ____________________

PROCEDURE (REMOVING GENERATOR)

1. Place fender covers on the fenders. Disconnect the negative (–) battery cable first and then the positive (+) battery cable at the battery. Task Completed ☐

   WARNING: *Never attempt to remove the generator without isolating the battery first.*

2. Disconnect the wiring leads from the generator. Task Completed ☐
3. Following the appropriate procedures, loosen the adjusting bolts and move the generator to provide sufficient slack to remove the generator drive belts. Task Completed ☐
4. Remove the through-bolts that retain the generator. Task Completed ☐
5. Remove the generator from the vehicle. Task Completed ☐

PROCEDURE (INSTALLING GENERATOR)

1. Install the generator onto its mounting bracket with bolts, washers, and nuts. Do not tighten. Task Completed ☐
2. Install the generator drive belts and tighten them to the proper belt tension. Task Completed ☐

   Specification ______________________________

3. Tighten all bolts to the specified torque. Task Completed ☐

   Specification ______________________________

4. Install the generator terminal plug and battery leads to the unit. Task Completed ☐
5. Connect the negative battery cable. Task Completed ☐
6. Check the generator for specified output and record: Task Completed ☐

   Specification ______________________________

   Actual ______________________________

   If the output is okay, remove the fender covers. Task Completed ☐

Problems Encountered

____

____

____

Instructor's Comments

____

____

____

ELECTRICAL AND ELECTRONIC SYSTEMS JOB SHEET 34

AC Generator Disassembly and Assembly

Name ______________________ Station ______________ Date ______________

NATEF Correlation

This Job Sheet addresses the following NATEF task:

D.4. Remove, inspect, and install generator (alternator).

Objective

Upon completion of this job sheet, you will be able to disassemble a generator (alternator) as well as clean, inspect, and test the components.

Tools and Materials

400-grain polishing cloth
Appropriate wrenches and screwdrivers
Clean rags
Cleaning solvent
Fender cover
High-temperature bearing grease
Press
Pulley puller
Screwdrivers
Scriber
Service manual
Small pry bar

Protective Clothing

Goggles or safety glasses with side shields

Describe the type of vehicle the generator was on:

Year ________ Make ____________________ Model ____________________

VIN ____________________ Engine type and size ____________________

Describe the type of generator:

____

PROCEDURE (DISASSEMBLY)

1. Place fender covers on the fenders. Disconnect the negative battery cable. Task Completed ☐

   WARNING: *Never attempt to disconnect the generator harness without first disconnecting the battery.*

2. Remove the generator from the vehicle. Task Completed ☐
3. Following the appropriate procedure, remove the brush assembly. Task Completed ☐
4. Remove the pulley nut. Using a suitable puller, remove the pulley. Task Completed ☐

   CAUTION: *Do not attempt to remove the pulley by prying it off.*

5. Remove the through-bolts. Mark the housing halves with a scriber. Pry the slip ring end frame and drive end frame apart, using a suitable small pry bar. Separate the units. Make sure the stator is pulled with the rear half. Task Completed ☐

PROCEDURE (ASSEMBLY)

1. Assemble in the reverse order of disassembly. Work carefully. Make sure all parts are correctly located. Align the scribe marks. If press fit is required, use a press. Task Completed ☐
2. Torque the pulley retaining nut. Do not clamp the rotor to hold it while torquing the pulley nut. Doing so may deform the rotor. Clamp the pulley. Task Completed ☐
3. Reassemble the brush unit. Some generator brush assembly designs permit the use of a pin to hold the brushes in the holder during assembly. Some require hooking the brush leads over a section of the brush holder for installation. Another type uses simple brush holders that permit installation after the generator is assembled. If a brush holder pin is used, or if the leads are hooked, remove the pin or straighten them after assembly. Task Completed ☐
4. Spin the rotor to check for free operation. Task Completed ☐
5. When mounting the generator to the engine, adjust the belt tension by prying on the edge of the heavy drive end frame. Never pry against the center or at the slip ring frame end. Tighten the belt to its tension specification and torque the bolts to specification. Task Completed ☐
6. Connect all wires to the generator. Connect the positive (+) battery cable first, then the negative (–) cable. Task Completed ☐
7. Remove the fender cover. Start the engine and test the charging system. Task Completed ☐

Problems Encountered

___

___

___

Instructor's Comments

___

___

___

ELECTRICAL AND ELECTRONIC SYSTEMS JOB SHEET 35

Diagnosing Light Circuits

Name ______________________ Station ______________ Date ______________

NATEF Correlation

This Job Sheet addresses the following NATEF tasks:

E.1. Diagnose the cause of brighter than normal, intermittent, dim, or no light operation; determine necessary action.

E.3. Inspect and diagnose incorrect turn signal or hazard light operation; perform necessary action.

Objective

Upon completion of this job sheet, you will be able to diagnose the cause of a brighter than normal, intermittent, dim, or no light operation as well as be able to inspect and diagnose turn signal or hazard light systems.

Tools and Materials

Wiring diagram for the vehicle

Digital multimeter (DMM)

Protective Clothing

Goggles or safety glasses with side shields

Describe the vehicle being worked on:

Year ________ Make ____________________ Model ____________________

VIN ____________________ Engine type and size ____________________

PROCEDURE

NOTE: *This job sheet focuses on the turn signal and hazard light circuits. The logic and techniques used to determine the cause of various problems can be applied to all light circuits. Simply identify the parts of the circuit you need to test and match their function to the components below.*

1. Operate the turn signals for the right side of the vehicle and describe what happened.

____

____

2. Operate the turn signals for the left side of the vehicle and describe what happened.

____

____

3. Operate the hazard lights on the vehicle and describe what happened.

____

____

4. Use the results of the above checks and match the problems with the following to diagnose the problem.

The bulbs burn brighter than normal

- Check the other lights, to see if they also burn brighter than normal. Check the voltage output of the charging system. If it is just the turn signals or hazard lights that burn brighter, check that circuit for a short. Summarize what you found.

____

____

The bulbs work intermittently

- If this problem only affects one bulb, check for a loose connection at that bulb or in that bulb's circuit. If only one side is affected, check for loose connections at the common points for those bulbs. If all bulbs are affected, check for a loose connection at the points that are common to all bulbs. Summarize what you found.

____

____

The bulbs are dimmer than normal

- If this problem only affects one bulb, check for high resistance in the power and ground circuits for that bulb. If only one side is affected, check for high resistance at the common points for those bulbs. If all bulbs are affected, check for high resistance at the points that are common to all bulbs. Summarize what you found.

____

____

None of the lamps light

- Check the fuse for that circuit. If the fuse is good, check for voltage at the common points in the circuit for all bulbs. Summarize what you found.

____

____

The hazards don't flash

- Check the fuse or circuit breaker. If the fuse is good, suspect the hazard flasher unit and substitute a known good one for the suspected flasher unit. Check for an open in the circuit that is common to all hazard lights; this includes the switch. Summarize what you found.

____

____

The turn signals light but don't flash

- Check the fuse or circuit breaker. If the fuse is good, suspect the turn signal flasher unit and substitute a known good one for the suspected flasher unit. Check for an open in the circuit that is common to all turn signal lights; this includes the switch and the circuit ground. Summarize what you found.

____

____

The turn signals flash at the wrong speed

- Suspect the turn signal flasher unit and substitute a known good one for the suspected flasher unit. If the flasher was good, check the available voltage at the flasher unit. If it is too high, check the voltage output of the charging system. If it is too low, check for high resistance in the power circuit to the flasher. If the voltage is normal, check the operation of all bulbs. Summarize what you found.

____

____

The front turn signals don't light

- Check for a loose connector or an open in the circuit that is only common to the front lights. Summarize what you found.

____

____

The rear turn signals don't light

- Check for a loose connector or an open in the circuit that is only common to the rear lights. Summarize what you found.

____

____

One turn signal bulb doesn't light

- Check the bulb. If the bulb is okay, check that lamp's circuit for an open or high resistance. Include a check of the ground circuit and of the bulb's socket. Summarize what you found.

Problems Encountered

Instructor's Comments

ELECTRICAL AND ELECTRONIC SYSTEMS JOB SHEET 36

Headlight Aiming

Name ______________________ Station ______________ Date ______________

NATEF Correlation

This Job Sheet addresses the following NATEF task:

E.2. Inspect, replace, and aim headlights and bulbs.

Objective

Upon completion of this job sheet, you will be able to inspect, replace, and aim headlights and bulbs.

Tools and Materials

A vehicle with adjustable headlights

Portable headlight aiming kit

Hand tools

Protective Clothing

Goggles or safety glasses with side shields

Describe the vehicle being worked on:

Year ________ Make ____________________ Model ____________________

VIN ____________________ Engine type and size ____________________

PROCEDURE

1. Describe the type of headlights used on the vehicle.

   ____

   ____

2. Park the vehicle on a level floor. Task Completed ☐

   Install the calibrated aiming units to headlights. Make sure the adapters fit the headlight aiming pads on the lens.

3. Zero the horizontal adjustment dial. Are the split image target lines visible in the view port?______

   If the lines cannot be seen, what should you do?

   ____

4. Turn the headlight horizontal adjusting screw until the split image target lines are aligned. Then repeat this for the other headlight. List any problems you may have had doing this.

____

____

5. Turn the vertical adjustment dial on the aiming unit to zero. Task Completed ☐

   Turn the vertical adjustment screw until the spirit level bubble is centered.

   Recheck your horizontal setting after adjusting the vertical.

6. List any problems you had making the vertical adjustment.

____

____

____

____

7. If the headlight assembly has four lamp assemblies, repeat steps 2 through 5 on the other two lamps. List below any problems you may have had doing this.

____

____

____

____

Problems Encountered

____

____

____

Instructor's Comments

____

____

____

ELECTRICAL AND ELECTRONIC SYSTEMS JOB SHEET 37

Working with High Intensity Discharge Headlights

Name ______________________ Station ______________ Date ______________

NATEF Correlation:

This Job Sheet addresses the following NATEF task:

E.4. Identify system voltage and safety precautions associated with high intensity discharge headlights.

Objective

Upon completion of this job sheet, you will be able to safely work around and on high intensity discharge (HID) headlights by being aware of the special precautions that must be adhered to.

Tools and Materials

Service manual

Protective Clothing

Goggles or safety glasses with side shields

Describe the vehicle being worked on:

Year ________ Make ____________________ Model ____________________

VIN ____________________ Engine type and size ____________________

PROCEDURE

1. How can you identify the high voltage circuit in this HID system?

   ____

   ____

2. What should you do if a HID headlight has cracks, dents, chips, or other damage?

   ____

   ____

3. Does the HID system on this vehicle have a fail-safe mode? If so, what does it do?

   ____

   ____

4. How many volts may be present at the high voltage socket for the headlamp when the lamp is turned on?

5. What is inside the bulb and why is it important to be aware of the bulb's construction?

6. When servicing a HID headlight, what things should be done to avoid electrical shock?

7. Why should you never turn on the headlights when there is no bulb in the headlamp?

8. Are there any other precautions given by the manufacturer that should be followed when servicing a HID? If so, what are they?

Problems Encountered

Instructor's Comments

ELECTRICAL AND ELECTRONIC SYSTEMS JOB SHEET 38

Bench Testing a Fuel Level Sending Unit

Name ______________________ Station ______________ Date ______________

NATEF Correlation

This Job Sheet addresses the following NATEF task:

F.1. Inspect and test gauges and gauge sending units for cause of intermittent, high, low, or no gauge readings; determine necessary action.

Objective

Upon completion of this job sheet, you will be able to inspect and test gauges and gauge sending units for cause of intermittent, high, low, or no gauge readings.

Tools and Materials

Digital multimeter (DMM)
Jumper wires
Service manual

Protective Clothing

Goggles or safety glasses with side shields

Describe the vehicle being worked on:

Year ________ Make __________________ Model __________________

VIN __________________ Engine type and size __________________

PROCEDURE

1. Locate the specifications for the fuel gauge-sending unit in the service manual. Record those specifications:

   ____

   ____

2. Select the ohmmeter function of the DMM. Task Completed ☐
3. Connect the DMM's negative test lead to the ground terminal of the sending unit. Task Completed ☐
4. Connect the meter's positive lead to the variable resistor terminal. Task Completed ☐
5. Holding the sender unit in its normal position, place the float rod against the empty stop. Task Completed ☐
6. Read the ohmmeter. What was the measurement?

   ____

7. Compare that measurement to the specifications. Your conclusions:

____

____

____

8. Slowly move the float toward the full stop, while observing the ohmmeter. The resistance change should be smooth and consistent. Your measurement was: _________ What does this indicate?

____

____

9. Check the resistance value while holding the float against the full stop. What was the measurement?

____

10. Compare that measurement to the specifications. Your conclusions:

____

____

____

11. Check the float to be sure that it is not filled with fuel, distorted, or loose. Record your findings:

____

____

12. Summarize the results of this test.

____

____

____

____

Problems Encountered

____

____

____

Instructor's Comments

____

____

____

ELECTRICAL AND ELECTRONIC SYSTEMS JOB SHEET 39

Checking Printed Circuits

Name ______________________ Station ______________ Date ______________

NATEF Correlation

This Job Sheet addresses the following NATEF task:

F.2. Inspect and test connectors, wires, and printed circuit boards of gauge circuits; determine necessary action.

Objective

Upon completion of this job sheet, you will be able to inspect and test connectors, wires, and printed circuit boards of gauge circuits.

Tools and Materials

Vehicle with a flexible circuit board for the instrument panel

Pencil with an eraser

Digital multimeter (DMM)

Service manual

Protective Clothing

Goggles or safety glasses with side shields

Describe the vehicle being worked on:

Year ________ Make ____________________ Model ____________________

VIN ____________________ Engine type and size ____________________

PROCEDURE

1. Remove the instrument cluster from the vehicle's dash. Task Completed ☐
2. Being careful not to touch the surface of the circuit board, visually inspect the printed circuit board for signs of tears and other damage. Record your findings.

   ____

   ____
3. Check the condition of the terminals for the edge connector to the circuit board. Record your findings.

   ____

   ____

4. Carefully clean the contacts for the edge connector by lightly rubbing them with the pencil's eraser. Task Completed ☐

5. Carefully remove the light bulbs from the circuit board and check them. Record your findings.

____

____

____

____

6. Reinstall the bulbs. Be careful not to tear the surface of the circuit board while doing this. Task Completed ☐

7. Using the service manual, identify the circuits for each of the terminal contacts on the circuit board and summarize what you found here.

____

____

____

8. With a high-impedance ohmmeter, check the continuity across each of those circuits. Record your findings.

____

____

9. Summarize your service recommendations for this circuit board.

____

____

____

____

Problems Encountered

____

____

____

Instructor's Comments

____

____

____

ELECTRICAL AND ELECTRONIC SYSTEMS JOB SHEET 40

Checking the Seat Belt Warning Circuitry

Name ______________________ Station ______________ Date ______________

NATEF Correlation

This Job Sheet addresses the following NATEF task:

F.3. Diagnose the cause of incorrect operation of warning devices and other driver information systems; determine necessary action.

Objective

Upon completion of this job sheet, you will be able to diagnose the cause of incorrect operation of warning devices and other driver information systems.

Tools and Materials

A vehicle with seat belts
Wiring diagram for the above vehicle
Owner's manual for the above vehicle
Digital multimeter (DMM)

Protective Clothing

Goggles or safety glasses with side shields

Describe the vehicle being worked on:

Year ________ Make ____________________ Model ____________________

VIN ____________________ Engine type and size ____________________

PROCEDURE

1. Describe the conditions during which the seat belt warning light and/or sound should be activated. Refer to the owner's manual.

   ____

   ____

2. Draw the seat belt warning circuit below. Identify the various controls of the circuit.

3. List all possible causes for an inoperative warning system. Base this on the wiring diagram and the owner's manual.

____

____

____

____

4. Test the operation of the system by opening and closing each of the system controls. Record your observations and conclusions from these tests below.

____

____

____

____

____

Problems Encountered

____

____

____

Instructor's Comments

____

____

____

ELECTRICAL AND ELECTRONIC SYSTEMS JOB SHEET 41

Checking Electronic Instrument Circuits

Name ______________________ Station ______________ Date ______________

NATEF Correlation

This Job Sheet addresses the following NATEF task:

F.4. Inspect and test sensors, connectors, and wires of electronic (digital) instrument circuits; determine necessary action.

Objective

Upon completion of this job sheet, you will be able to inspect and test sensors, connectors, and wires of electronic instrument circuits.

Tools and Materials

A vehicle with accessible sensors and switches

Service manual for the vehicle

Component locator manual for the vehicle

A DSO

Digital multimeter (DMM)

Protective Clothing

Goggles or safety glasses with side shields

Describe the vehicle being worked on:

Year ________ Make ____________________ Model ____________________

VIN ____________________ Engine type and size ____________________

PROCEDURE

1. Refer to the service manual for the proper procedure for running self-diagnostics on the instrument cluster. Summarize the procedure for doing this.

   ____

   ____

   ____

2. Summarize your findings from running self-diagnostics.

   ____

   ____

   ____

3. What are your service recommendations?

____

____

____

4. If there were no trouble codes retrieved or all found faults were corrected, proceed to diagnose the system based on the following symptoms.

Gauge Reads Low Constantly

1. Disconnect the wire harness from the sending unit of the malfunctioning gauge. Task Completed ☐
2. Connect a jumper wire between the gauge and the sending unit. Task Completed ☐
3. Turn the ignition switch on and observe the gauge. Record what happened.

____

____

4. Explain why you connected the jumper wire. What were you looking for?

____

____

5. If the gauge reads too high with the jumper wire in place, visually inspect the ground for the sending unit and record your findings.

____

____

6. Check the ground with the DMM and record your findings.

____

____

7. If the ground was good and the gauge doesn't read correctly, what could be the cause?

____

____

8. If the gauge still reads low with the jumper wire, check the sending unit with an ohmmeter. Record your findings.

____

____

9. What are your service recommendations?

____

____

Gauge Reads High Constantly

Task Completed ☐

1. Disconnect the wire harness from the sending unit of the malfunctioning gauge.
2. Turn the ignition switch on and observe the gauge. Record what happened.

____

____

3. What is the cause of the problem if the gauge now reads low?

____

____

4. If the gauge reads too high with the harness disconnected, visually inspect the wire from the sending unit to the gauge connector and record your findings.

____

____

5. Check for a short to ground in that circuit with the DMM and record your findings.

____

____

6. What are your service recommendations?

____

____

____

Inaccurate Gauge Readings

1. Refer to the service manual and identify the resistance values for the sending unit of the malfunctioning gauge. Record those specifications.

____

____

2. Remove the sending unit and check its resistance. Summarize your findings.

____

____

____

3. Compare your measurements to the specifications and state what your service recommendations are.

____

____

4. If the sending unit checked out fine, visually inspect all wires and connectors in the circuit and record your findings.

____

____

5. Check the wires and connectors with the DMM and record your findings.

____

____

6. What are your service recommendations?

____

____

____

Problems Encountered

____

____

____

Instructor's Comments

____

____

____

ELECTRICAL AND ELECTRONIC SYSTEMS JOB SHEET 42

Diagnosing Horn Problems

Name ______________________ Station ______________ Date ______________

NATEF Correlation

This Job Sheet addresses the following NATEF task:

G.1. Diagnose incorrect horn operation; perform necessary action.

Objective

Upon completion of this job sheet, you will be able to diagnose incorrect horn operation.

Tools and Materials

Jumper wires
Digital multimeter (DMM)
Test light
Hand tools
Service manual

Protective Clothing

Goggles or safety glasses with side shields

Describe the vehicle being worked on:

Year ________ Make ____________________ Model ____________________

VIN ____________________ Engine type and size ____________________

PROCEDURE

NOTE: *Diagnosis of the horn circuit begins with the problem or customer's complaint. The procedures in this job sheet are divided by symptom. Choose the symptom that best describes the customer's complaint and follow that procedure. Although not all of the symptoms apply to one problem, it is wise to look through the procedures for all symptoms.*

Horn Doesn't Work

1. Verify the customer's complaint by depressing the horn button. Record your findings.

   ____

   ____

2. Rotate the steering wheel from stop to stop while depressing the horn button. Did the horn work? What are you checking for?

   ____

   ____

3. If the horn didn't work, check the fuse or fusible link. Describe its condition.

____

____

4. Connect a jumper wire from the positive terminal of the battery to the horn terminal. Did the horn sound? ________ What does this indicate?

____

____

5. If the horn didn't sound, check the ground at the horn and record your findings.

____

____

6. Summarize your service recommendations.

____

____

____

Poor Sound

1. Does the horn system have more than one horn? If yes, do both horns turn on when the horn button is depressed?

____

____

2. If one horn is not working, that will affect the overall quality of the horn sound. If this is the case, diagnose the cause of the inoperative horn and record your results.

____

____

3. Use a voltmeter to measure the voltage to the horn. What did you find?

____

____

4. If the voltage is less than what you would expect, check the voltage drops on the power side and ground side of the circuit. What did you find?

____

____

5. If no electrical problems were found, connect a jumper wire from the battery to the horn. Task Completed ☐

6. Turn out the adjusting screw at the horn approximately one-quarter turn. Did the sound improve?

___

___

7. Adjust the screw as needed if the screw made a difference in sound quality. If the screw had no effect, replace the horn. Task Completed ☐

Horn Sounds Continuously

1. Disconnect the horn relay from the circuit.
2. With an ohmmeter, check for continuity from the battery terminal of the relay to the horn circuit terminal. What is indicated if there is continuity?

___

___

3. With an ohmmeter check the continuity through the horn switch. What is indicated if there is continuity?

___

___

4. What are your service recommendations?

___

___

___

Problems Encountered

___

___

___

Instructor's Comments

___

___

___

ELECTRICAL AND ELECTRONIC SYSTEMS JOB SHEET 43

Diagnosing a Windshield Wiper Circuit

Name ______________________ Station ______________ Date ______________

NATEF Correlation

This Job Sheet addresses the following NATEF task:

G.2. Diagnose incorrect wiper operation; diagnose wiper speed control and park problems; perform necessary action.

Objective

Upon completion of this job sheet, you will be able to diagnose incorrect wiper operation and diagnose wiper speed control and park problems.

Tools and Materials

Wiring diagram for the vehicle

Service manual for the vehicle

Digital multimeter (DMM)

Protective Clothing

Goggles or safety glasses with side shields

Describe the vehicle being worked on:

Year ________ Make ____________________ Model ____________________

VIN ____________________ Engine type and size ____________________

PROCEDURE

1. Describe the general operation of the windshield wipers. Check the operation in all speeds and modes.

   ____

   ____

   ____

2. Check the mechanical linkages for evidence of binding or breakage. Record your findings.

   ____

   ____

   ____

3. Draw the windshield wiper circuit below. Include its power source, controls, and ground.

4. Describe how the motor is controlled to operate at different speeds.

____

____

____

____

5. Connect the voltmeter across the ground circuit, and then energize the motor. What was your reading on the meter? What does this indicate?

____

____

____

____

6. Probe the power feed to the motor in the various switch positions (including the OFF or PARK position). Observe your voltmeter readings. What were they? What is indicated by these readings?

____

____

____

____

7. Describe the general operation of the windshield wiper system.

____

____

____

____

Problems Encountered

Instructor's Comments

ELECTRICAL AND ELECTRONIC SYSTEMS JOB SHEET 44

Diagnosing Windshield Washer Problems

Name ______________________ Station ______________ Date ______________

NATEF Correlation

This Job Sheet addresses the following NATEF task:

G.3. Diagnose incorrect windshield washer operation; perform necessary action.

Objective

Upon completion of this job sheet, you will be able to diagnose incorrect windshield washer operation.

Tools and Materials

Clean rag
Catch basin
Digital multimeter (DMM)
Service manual

Protective Clothing

Goggles or safety glasses with side shields

Describe the vehicle being worked on:

Year ________ Make ____________________ Model ____________________

VIN ____________________ Engine type and size ____________________

PROCEDURE

1. Check the level of the fluid in the washer fluid container. Fill the container to the correct level, if necessary. What did you find?

 ____

 ____

2. Before beginning your diagnosis of the washer system, activate it and observe the sounds and actions of the system when you turn it on. Describe what you observed.

 ____

 ____

3. If the motor seems to run, check the lines for restrictions by removing the hose from the pump. Task Completed ☐

4. Activate the system and describe what happened.

 ____

 ____

5. What are your service recommendations so far?

____

____

____

6. If you determined there is a restriction somewhere in the fluid delivery or spray circuit, what can you do to identify the cause of the problem?

____

____

7. If the pump did not push out fluid or if the motor did not seem to run during the initial check, check the fuse and record its condition.

____

____

8. Activate the pump and look for signs that it is operating. What did you find?

____

____

9. If the pump works, check the fluid feed line for restrictions or damage. Record your results.

____

____

10. If the pump doesn't work, use a voltmeter to check for available voltage at the pump when it is activated. What did you find?

____

____

11. If there is power to the pump, check the pump's ground circuit. What did you find?

____

____

12. If the ground is good and there is power to the pump, the pump is bad and should be replaced. Task Completed ☐

13. If there is no power to the pump, check for power to the switch. What did you find?

____

____

14. If there is power to the switch, check for power out of the switch. What did you find?

____

____

15. If there is power to the switch but no power out of the switch, replace the switch. Task Completed ☐

16. If there is power in and out of the switch, check the power feed circuit to the pump. What did you find and what is the cause of the problem?

____

____

Problems Encountered

____

____

____

Instructor's Comments

____

____

____

ELECTRICAL AND ELECTRONIC SYSTEMS JOB SHEET 45

Diagnosing Motor-Driven Accessories

Name ______________________ Station ______________ Date ______________

NATEF Correlation

This Job Sheet addresses the following NATEF task:

H.1. Diagnose incorrect operation of motor-driven accessory circuits; determine necessary action.

Objective

Upon completion of this job sheet, you will be able to inspect and test A/C-heater blower motor circuits and components.

Tools and Materials

Wiring diagram for the vehicle

Digital multimeter (DMM)

Protective Clothing

Goggles or safety glasses with side shields

Describe the vehicle being worked on:

Year ________ Make ____________________ Model ____________________

VIN ____________________ Engine type and size ____________________

PROCEDURE

NOTE: *Although this job sheet is focused on the blower motor, the procedures can be easily applied to all motor-driven accessories. Refer to the appropriate service manual and wiring diagram to identify the components of the circuit you wish to test. Then apply the same sequence and logic given in this job sheet.*

1. Refer to the service or owner's manual to identify the number of speeds the blower motor should operate at. Also, identify whether the motor should run when the switch is in the off or its lowest position. Describe how the blower motor should operate.

   ____

   ____

2. Check the wiring diagram and identify whether the blower circuit is controlled by a ground side switch or an insulated (power side) switch. State which one.

   ____

   ____

3. Start the engine and turn on the blower. Move the blower control to all available positions and summarize what happened in each.

____

____

4. Match the type of circuit control and the problem with one of the following.

The blower works at some speeds but not all

a. Check the voltage to the blower motor at the various switch positions.
b. If the voltage doesn't change when a new position is selected, check the circuit from the switch to the resistor block.
c. If there was zero voltage in a switch position, check for an open in the resistor block.
d. Give a summary of the test results.

____

____

The blower doesn't work and is controlled by an insulated switch

a. Check the fuse or circuit breaker. If it is open, check the circuit for a short.
b. Connect a jumper wire from a power source to the motor. If the motor operates, there is an open in the circuit between the fuse and the motor.
c. To identify the location of the open, connect a jumper wire across the switch. If the motor now operates, the switch is bad.
d. Then, connect a jumper wire across the resistor block. If the motor now operates, the resistor block is bad.
e. If the motor did not operate when the jumper wire connected it to a power source, connect the jumper wire from the motor to a known good ground. If the motor operates, the fault is in the ground circuit. If the motor doesn't operate, the motor is bad.

f. Give a summary of the test results.

____

____

The blower doesn't work and is controlled by a ground side switch

a. Check the fuse or circuit breaker. If it is open, check the circuit for a short.
b. Connect a jumper wire from the motor's ground terminal to a known good ground. If the motor runs, the problem is in that circuit.
c. If the motor did not operate, check for voltage at the battery terminal of the motor. If no voltage is found, there is an open in the power feed circuit for the motor. If voltage was present, the motor is bad.
d. If the motor operated when the jumper wire was connected in step **b**, use an ohmmeter to check the ground connection of the switch.

e. If the ground is good, use a voltmeter to probe for voltage at any of the circuits from the resistor block to the switch. Replace the switch if there is power at that point.
f. If no voltage was available at the switch, check the circuit for an open. If there is not an open, suspect a faulty switch.
g. Give a summary of the test results.

____

____

Constantly running blower fan

a. If the motor runs when it should not, there is probably a short in the circuit. If a ground side switch controls the circuit, check for a short to ground in the control circuit. The exact problem can be isolated by disconnecting portions of the circuit until the motor stops. The short is in that part of the circuit that disconnected last.
b. If the circuit is controlled by an insulated switch, check for a wire-to-wire short. Check other circuits of the vehicle to identify what circuit is involved in this problem. That circuit will also experience a lack of control or when that circuit is turned off, the blower motor will turn off. The exact problem can be isolated by disconnecting portions of the circuit until the motor stops. The short is in that part of the circuit that disconnected last.
c. Give a summary of the test results.

____

____

Problems Encountered

____

____

____

Instructor's Comments

____

____

____

ELECTRICAL AND ELECTRONIC SYSTEMS JOB SHEET 46

Diagnosing Heated Accessories

Name ______________________ Station ______________ Date ______________

NATEF Correlation:

This Job Sheet addresses the following NATEF task:

H.2. Diagnose incorrect heated glass, mirror, or seat operation; determine necessary action.

Objective

Upon completion of this job sheet, you will be able to accurately check the operation of various heated accessories and determine what is needed to repair them

Tools and Materials

Service manual
Test light
Digital multimeter (DMM)
Soft, dry cloth
Heat lamp
Thermometer

Protective Clothing

Goggles or safety glasses with side shields

Describe the vehicle being worked on:

Year ________ Make ____________________ Model ____________________

VIN ____________________ Engine type and size ____________________

PROCEDURE

Rear Window Defogger

1. Refer to the service manual and briefly describe where the main components for this circuit are.

____

____

2. To check the defogger relay, use an ohmmeter check the continuity across the various terminals of the relay. The service manual will give these and the expected test results. What are the specifications for the various terminals of the relay?

____

____

3. What were your measurements and what is indicated by them?

____

____

4. If the readings were outside of specifications, what should be done?

____

____

5. Wipe the inside of the back window with a soft, dry cloth, and be sure to wipe the glass in the direction of the defogger wires. Try not to damage the wires. Do not use detergents or glass cleaners with abrasive ingredients. Task Completed ☐

6. Turn the ignition switch ON. Task Completed ☐

7. Turn the defogger switch ON. Task Completed ☐

8. Check for voltage at the center of each heat wire, did you have about 5 volts on each wire? What is indicated by your measurement?

____

____

9. If a break in the wire is indicated, place the voltmeter positive lead on the defogger's positive terminal. Then, place the voltmeter negative lead on the heat wire at the positive terminal end and slide it toward the negative terminal end. The point where the voltmeter deflects from zero to several volts is the place where the heat wire is broken. What did you find?

____

____

10. If necessary, repair the defogger wire by cleaning the broken wire tips with a grease, wax, and silicone remover. Task Completed ☐

11. Place the masking tape along both sides of the wire to be repaired. Task Completed ☐

12. Thoroughly mix the repair agent and with a fine tip brush, apply a small amount to the wire. Task Completed ☐

13. After a few minutes, remove the masking tape. Allow the repair to stand at least 24 hours before using it. Task Completed ☐

Outside Mirror Heater

1. Refer to the service manual and briefly describe where the main components for this circuit are.

____

____

2. With the mirror circuit turned on (on some vehicles, this turns on with the rear window defogger), feel the mirror to see if it is heating up. Typically, it takes awhile for the mirror to become warm. What did you find?

____

____

3. If the mirror did not heat or if seems that it took too long, check the voltage drop across the mirror. If battery voltage is dropped across it, the mirror is fine. If the voltage drop is lower than expected, check both the power and ground circuits of the mirror for excessive resistance or other problem. What were your measurements and what is indicated by them?

____

____

4. If the readings were outside of specifications, what should be done?

____

____

Heated Seats

1. Refer to the service manual and briefly describe where the main components for this circuit are.

____

____

2. If the seats do not heat, begin by checking the continuity of the seat heater switch. Follow the instructions for doing this as given in the service manual. Record your test results here:

____

____

3. What is indicated by your measurements?

____

____

4. Locate the connector to the seat cushion heater and check for continuity through it under the conditions outlined by the service manual. Normally these measurements are taken with heat applied to the thermostat in the seat. Use the heat lamp to bring the thermostat to the specified temperatures. Record your test results here:

____

____

5. What is indicated by your measurements?

____

____

6. Locate the connector to the seat back heater and check for continuity through it under the conditions outlined by the service manual. Record your test results here:

____

____

7. What is indicated by your measurements?

____

____

Problems Encountered

____

____

____

Instructor's Comments

____

____

____

ELECTRICAL AND ELECTRONIC SYSTEMS JOB SHEET 47

Diagnosing Power Lock Circuits

Name ______________________ Station ______________ Date ______________

NATEF Correlation

This Job Sheet addresses the following NATEF task:

H.3. Diagnose incorrect electric lock operation; determine necessary action.

Objective

Upon completion of this job sheet, you will be able to diagnose incorrect electric lock operation.

Tools and Materials

Jumper wires

Digital multimeter (DMM)

Service manual

Protective Clothing

Goggles or safety glasses with side shields

Describe the vehicle being worked on:

Year ________ Make ____________________ Model ____________________

VIN ____________________ Engine type and size ____________________

PROCEDURE

1. If one door lock doesn't work, apply battery voltage to the motor and describe what happened.

   ____

   ____

2. Reverse the connection at the motor and describe what happened.

   ____

   ____

3. If the motor didn't work in both or either connections, replace the motor. Task Completed ☐

4. If the motor worked, remove the switch for that door. Task Completed ☐

5. Check the continuity across the switch in all positions and record your findings.

   ____

   ____

6. If the door switch is okay or if all of the door locks don't work, remove the master switch. Task Completed ☐

7. Check the continuity across the switch in all positions and record your findings.

____

____

8. Summarize your service recommendations.

____

____

____

Problems Encountered

____

____

____

Instructor's Comments

____

____

____

ELECTRICAL AND ELECTRONIC SYSTEMS JOB SHEET 48

Speed or Cruise Control Simulated Road Test

Name ______________________ Station ______________ Date ______________

NATEF Correlation

This Job Sheet addresses the following NATEF task:

H.4. Diagnose incorrect operation of cruise control systems; repair as needed.

Objective

Upon completion of this job sheet, you will be able to diagnose the incorrect operation of cruise control systems.

Tools and Materials

Hoist

Service manual

Two jack stands

Two wood blocks or wedges

Protective Clothing

Goggles or safety glasses with side shields

Describe the vehicle being worked on:

Year ________ Make ____________________ Model ____________________

VIN ____________________ Engine type and size ____________________

WARNING: *This procedure should only be conducted when your instructor is readily available to supervise you.*

PROCEDURE

CAUTION: *When performing the simulated road test, the drive wheels of the vehicle must be raised clear of the floor. Block the wheels securely and use only a suitable lifting device (such as a garage-type hoist). If a rear-wheel-drive vehicle is being tested, support the rear axle with one jack stand on each side. Never attempt to use the vehicle bumper jack for tests of this type.*

1. Perform visual inspection of speed control system. If all right, raise the vehicle on hoist. Task Completed ☐
2. Start the engine. Shift the transmission into DRIVE. Task Completed ☐
3. Turn on speed control. Task Completed ☐

CAUTION: *If at any time during the following steps the system should appear to go out of control and overspeed, be prepared to turn the system off at once.*

4. Accelerate and hold at 35 mph. Press and release the SET ACCEL button. Hold foot pressure very lightly on the accelerator pedal. Normally speed will continue at 35 mph for a short period of time and then gradually start surging because the engine is not loaded. Record your results on the Report Sheet for Speed Control Simulated Road Test, found at end of this job sheet. Task Completed ☐

5. Press the OFF button. The engine should drop back to idle. Use the brakes to stop the drive wheels. Record your results on the Report Sheet for Speed Control Simulated Road Test. Task Completed ☐

6. Press the ON button, accelerate, and hold the speed at 35 mph. Press and hold the SET ACCEL button. Slowly remove your foot from the accelerator. The engine speed should gradually increase. Record your results on the Report Sheet for Speed Control Simulated Road Test. Task Completed ☐

7. When the speed reaches 50 mph, release the SET ACCEL button. The vehicle should maintain 50 mph for a short time before the surging begins. Record your results on the Report Sheet for Speed Control Simulated Road Test. Task Completed ☐

   CAUTION: *Do not allow the vehicle to run at 50 mph for an extended time. The axle speed is actually 100 mph and may cause differential damage.*

8. Press the COAST button and hold. The engine should idle. Slow the drive wheels to 35 mph. The vehicle should begin surging. Record your results on the Report Sheet for Speed Control Simulated Road Test. Task Completed ☐

9. Press and release the brake pedal. The system should shut off and the engine should slow to idle. Record your results on the Report Sheet for Speed Control Simulated Road Test. Task Completed ☐

10. Accelerate the engine and set the speed at 50 mph. Brake to 35 mph and maintain 35 mph with the accelerator. Depress and release the RESUME button. The speed should return to 50 mph. Record your results on the Report Sheet for Speed Control Simulated Road Test. Task Completed ☐

Problems Encountered

____

____

____

Instructor's Comments

____

____

____

Name ______________________ Station ______________ Date ______________

REPORT SHEET FOR SPEED CONTROL SIMULATED ROAD TEST		
	OK	*Service Required*
1. Visual inspection		
2. SET ACCEL	**Yes**	**No**
Hold at 35 mph		
Idle when OFF		
Increase from 35 mph		
Maintain 50 mph		
3. Set COAST		
Return to idle		
Maintain 35 mph		
4. Depress brake pedal		
Return to idle		
5. Set RESUME		
Return to 50 mph		

Conclusions and Recommendations ______________________________________

____

____

ELECTRICAL AND ELECTRONIC SYSTEMS JOB SHEET 49

Working Safely Around Air Bags

Name ______________________ Station ______________ Date ______________

NATEF Correlation

This Job Sheet addresses the following NATEF task:

H.5. Diagnose supplemental restraint system (SRS) concerns; determine necessary action. (Note: Follow manufacturer's safety procedures to prevent accidental deployment.)

Objective

Upon completion of this job sheet, you will be able to safely diagnose supplemental restraint system (SRS) concerns.

Tools and Materials

A vehicle with air bags

Service manual for the above vehicle

Component locator for the above vehicle

Safety glasses

Digital multimeter (DMM)

Protective Clothing

Goggles or safety glasses with side shields

Describe the vehicle being worked on:

Year ________ Make ________________ Model ________________

VIN ________________ Engine type and size ________________

List all of the restraint systems found on this vehicle:

____

____

____

PROCEDURE

1. Locate the information about the air bag system in the service manual. How are the critical parts of the system identified in the vehicle?

____

____

____

____

2. List the main components of the air bag system and describe their location.

____

____

____

____

____

____

____

____

3. Here are some very important guidelines to follow when working with and around air bag systems. These are listed below with some key words left out. Read through these and fill in the blanks with the correct words.

   a. Wear ________ __________ when servicing an air bag system and when handling an air bag module.
   b. Wait at least _______ minutes after disconnecting the battery before beginning any service. The reserve _________ module is capable of storing enough energy to deploy the air bag for up to ________ minutes after battery voltage is lost.
   c. Always handle all ________ and other components with extreme care. Never strike or jar a sensor, especially when the battery is connected. Doing so can cause deployment of the air bag.
   d. Never carry an air bag module by its __________ or __________, and, when carrying it, always face the trimmed side of the module _______ from your body. When placing a module on a bench, always face the trimmed side of the module ______.
   e. Deployed air bags may have a powdery residue on them. __________ _________ is produced by the deployment reaction and is converted to ____________ ____________ when it comes in contact with the moisture in the atmosphere. Although it is unlikely that harmful chemicals will still be on the bag, it is wise to wear _______ _________ and _________ when handling a deployed air bag. Immediately wash your hands after handling a deployed air bag.
   f. A live air bag must be _____________ before it is disposed. A deployed air bag should be disposed of in a manner consistent with the __________ and manufacturer's procedures.
   g. Never use a battery- or AC-powered ___________, _____________, or any other type of test equipment in the system unless the manufacturer specifically says to. Never probe with a ______ ______ for voltage.

Problems Encountered

____

____

____

Instructor's Comments

____

____

____

ELECTRICAL AND ELECTRONIC SYSTEMS JOB SHEET 50

Disarming and Enabling an Air Bag System

Name ______________________ Station ______________ Date ______________

NATEF Correlation

This Job Sheet addresses the following NATEF task:

H.6. Disarm and enable the air bag system for vehicle service.

Objective

Upon completion of this job sheet, you will be able to disarm and enable an air bag system so you can safely work around and near the air bag without accidentally deploying it and making it safe for the driver after you have serviced the vehicle.

Tools and Materials

Wiring diagram for the vehicle

Protective Clothing

Goggles or safety glasses with side shields

Describe the vehicle being worked on:

Year ________ Make ____________________ Model ____________________

VIN ____________________ Engine type and size ____________________

PROCEDURE

NOTE: *This is a typical procedure. ALWAYS refer to the service manual for the exact procedure. If you are following the directions given in the service manual and they differ from the steps included in this worksheet, write the changes in procedure on the worksheet.*

1. Disconnect the negative battery cable. Task Completed ☐
2. Tape the terminal of the cable to prevent accidental contact with the post of the battery. Task Completed ☐
3. Remove the air bag system (SIR, SRS) fuse from the fuse block. Task Completed ☐
4. Wait at least 10 seconds before proceeding. Task Completed ☐
5. Disconnect the yellow connector at the base of the steering column. Task Completed ☐
6. After service to the air bag or related system, reconnect the yellow connector at the base of the steering column. Task Completed ☐
7. Install the fuse for the air bag system. Task Completed ☐
8. Reconnect the negative battery cable. Task Completed ☐

9. Perform a system self-test to make sure the system is enabled properly and ready for an emergency. How did you perform the self-test?

____

____

Problems Encountered

____

____

____

Instructor's Comments

____

____

____

ELECTRICAL AND ELECTRONIC SYSTEMS JOB SHEET 51

Identifying the Source of Static on a Radio

Name ______________________ Station ______________ Date ______________

NATEF Correlation

This Job Sheet addresses the following NATEF task:

H.7. Diagnose radio static and weak, intermittent, or no radio reception; determine necessary action.

Objective

Upon completion of this job sheet, you will be able to correctly diagnose radio static and weak, intermittent, or no radio reception.

Tools and Materials

Jumper wire with alligator clips on both ends

Protective Clothing

Goggles or safety glasses with side shields

Describe the vehicle being worked on:

Year ________ Make ____________________ Model ____________________

VIN ____________________ Engine type and size ____________________

Describe the sound system in the vehicle:

____

____

____

PROCEDURE

1. Turn on the radio and listen for the noise. Typically the noise is best heard on the low AM stations. Describe the noise:

   ____

   ____

2. Listen to the radio with the engine off and with it running. Describe the difference in the noise level and the reception of the radio.

   ____

   ____

3. What can you conclude from your answer to number 2?

____

____

4. Operate the radio in AM and FM. Does the noise appear in both bands?

____

5. If the noise is only on FM, what could be the problem?

____

____

6. If the noise is heard on both AM and FM, continue the test by checking the antenna and its connections. Is the antenna firmly mounted and in good condition?

____

____

7. Check the connection of the antenna cable to the antenna. Are the contacts clean and is the cable connector in good condition?

____

____

8. Connect a jumper wire from the base of the antenna to a known good ground. Then listen for the noise. Did the noise level change? Describe and explain the results.

____

____

9. Refer to the service manual and identify any noise suppression devices used on this vehicle. The noise suppression devices are:

____

____

10. Are all of the noise suppression devices present on the vehicle and are they mounted securely to a clean well-grounded surface?

____

____

11. Connect a jumper wire from a known good ground to the grounding tab on each capacitor-type noise suppressor. Listen to the radio and describe and explain the results of doing this:

____

____

12. Turn off the engine and disconnect the wiring harness from the voltage regulator to the generator. Start the engine and listen to the radio and describe and explain the results of doing this:

____

____

13. Check the spark plug wires and spark plugs. Are both of these noise suppressor-types?

____

14. Check the routing, condition, and connecting points for the spark plug cables. Describe their condition:

____

____

15. Connect a jumper wire from a known good ground to the frame of a rear speaker. Listen to the radio and describe and explain the results of doing this:

____

____

16. What are your conclusions and recommendations based on this job sheet?

____

____

____

____

17. If the noise source has not been identified, what should you do next?

____

____

Problems Encountered

____

____

____

Instructor's Comments

____

____

____

ELECTRICAL AND ELECTRONIC SYSTEMS JOB SHEET 52

Remove and Reinstall a Door Panel

Name ____________________ Station ______________ Date ______________

NATEF Correlation

This Job Sheet addresses the following NATEF task:

H.8. Remove and reinstall door panel.

Objective

Upon completion of this job sheet, you will be able to remove and reinstall an inner door panel. This may be necessary to gain access to the various electrical and mechanical systems that are housed within the door.

Tools and Materials

Hand tools
Service manual
Soft-faced hammer
Special tools for panel plug removal

Protective Clothing

Goggles or safety glasses with side shields

Describe the vehicle being worked on:

Year ________ Make ____________________ Model ____________________

VIN ____________________ Engine type and size ____________________

PROCEDURE

1. Install a power source so the electronic memories of the various accessories and computers can be maintained. Task Completed ☐
2. Disconnect the negative cable of the battery. Task Completed ☐
3. Locate any service precautions given in the service manual, especially those pertaining to air bags and electronics. List and describe all precautions.

____

____

____

____

4. Which door panel are you planning on removing?

____

____

5. Again referring to the service manual, summarize the procedure for removing the door panel you are about to remove.

____

____

____

____

6. Identify the components that will need to be removed in order to remove the door panel; include all speakers, switches, armrests, and compartments.

____

____

____

____

7. Begin your removal of the door panel by removing the screws and/or bolts that hold armrests and other major components to the panel. What did these retain?

____

____

8. Now examine all switches to determine if they should be removed once the panel is loose or prior to that. Describe your findings.

____

9. Remove all switches and everything else that may be directly attached to the panel. Task Completed ☐

10. Starting at one end and using the correct tool, disengage the panel retaining clips from the door, being careful not to pull too hard or to damage the clips or the panel. Task Completed ☐

11. Once the plugs are all free from their bores, lift the panel slightly out and up and check for anything that will interfere with the removal of the panel. What else needs to be removed?

____

____

12. Remove the panel and keep track of the retaining clips and plugs. Task Completed ☐

13. Replace any plugs and clips that may have been damaged. Task Completed ☐

14. Carefully position the panel so that the clips and plugs are over their respective bores, then press or lightly pound the panel with the clips into position. Task Completed ☐

15. Reinstall all switches and everything else that was directly attached to the panel and was removed. Task Completed ☐

16. Reinstall and tighten the screws and/or bolts that hold armrests and other major components to the panel. Task Completed ☐

17. Reconnect the negative cable of the battery. Task Completed ☐

18. Disconnect the electronic memory keeper. Task Completed ☐

Problems Encountered

___

___

___

Instructor's Comments

___

___

___

ELECTRICAL AND ELECTRONIC SYSTEMS JOB SHEET 53

Using a Scan Tool on Body Control Systems

Name ______________________ Station ______________ Date ______________

NATEF Correlation

This Job Sheet addresses the following NATEF tasks:

H.9. Diagnose body electronic system circuits using a scan tool; determine necessary action.

H.10. Check for module communication (LAN/CAN/BUS) errors using a scan tool.

Objective

Upon completion of this job sheet, you will be able to connect a scan tool to a body control system and retrieve trouble codes and check for communication errors.

Tools and Materials

Hand tools

Service manual

Scan tool

Protective Clothing

Goggles or safety glasses with side shields

Describe the vehicle being worked on:

Year ________ Make ____________________ Model ____________________

VIN ____________________ Engine type and size ____________________

PROCEDURE

1. Using the service manual as an information source, describe the systems and functions that are controlled by the body control module.

   ____

   ____

   ____

   ____

2. Check and record the voltage of the battery.

   ____

   ____

3. Is this voltage sufficient for proper operation of the body control system and electric accessories? Why or why not?

____

____

4. Visually inspect the sensors and switches of the body control system and summarize your findings.

____

____

5. Check all visible and accessible ground connections for integrity. Describe your findings here.

____

____

6. How would a bad ground affect the operation of a sensor and/or complete circuit?

____

____

7. Check all wiring for signs of burned or chaffed spots, pinched wires, or contact with sharp edges or hot exhaust parts. Describe your findings here.

____

____

8. Visually check all vacuum lines and hoses for integrity. Describe your findings.

____

____

9. Using the service manual as a guide, describe the procedure for retrieving trouble codes from the body control module.

____

____

10. Describe the method used to monitor the data stream from the body computer. What did you need to do with the scan tool?

____

____

____

____

11. How do you know when the components of the system are communicating properly with each other?

Problems Encountered

Instructor's Comments

ELECTRICAL AND ELECTRONIC SYSTEMS JOB SHEET 54

Diagnosing Anti-Theft Systems

Name ______________________ Station ______________ Date ______________

NATEF Correlation

This Job Sheet addresses the following NATEF task:

H.11. Diagnose the cause of false, intermittent, or no operation of anti-theft system.

Objective

Upon completion of this job sheet, you will be able to diagnose anti-theft systems.

Tools and Materials

Digital multimeter (DMM)

Hand tools

Service manual

Protective Clothing

Goggles or safety glasses with side shields

Describe the vehicle being worked on:

Year ________ Make ____________________ Model ____________________

VIN ____________________ Engine type and size ____________________

PROCEDURE

NOTE: *There are many different designs of anti-theft systems, each with its own diagnostic and service procedures. Make sure you follow the guidelines given by the manufacturer before proceeding. This job sheet is designed to take you through the typical steps required to identify the cause of common problems.*

A. System Will Not Disarm

1. Check the lock cylinder switches for looseness or damage and describe what you found.

____

____

2. Attempt to disable the system by inserting the key into the passenger door's lock cylinder. If the system disarms, what is indicated?

____

____

3. If the passenger side did not disarm the system, with the key, hold the lock cylinder in the unlock position and check for opens in the circuit between the cylinder switch and the controller. Describe what you found.

____

____

B. Alarm Goes Off by Itself

1. Using the wiring diagram, locate the diodes in the system. Describe where they are.

____

____

2. With your DMM check each diode and summarize what you found.

____

____

3. Carefully check all of the ground circuits and connections in the system. Summarize what you found.

____

____

4. Check the lock cylinder, tamper, and door jamb switches for looseness and damage. Record your findings.

____

____

C. Security Light Blinks

1. Check the tamper switches for looseness. Describe your findings.

____

____

2. Inspect the wires that connect the door jamb switches and the controller. Describe what you found.

____

____

3. Inspect the wires that connect the tamper jamb switches and the controller. Describe what you found.

____

____

D. Security Light Is On and Won't Disarm

1. Carefully inspect the door and trunk lock cylinders for evidence of tampering or damage. Record your findings.

____

____

2. Check the tamper switches for looseness and record your findings.

____

____

3. Inspect the wires at the door jamb switches and lock cylinders. Describe what you found.

____

____

4. Check the jamb switches for proper operation and adjustment. What did you find?

____

____

5. Carefully inspect all wires external to the controller and its relays. Describe what you found.

____

____

E. Security Lamp Is Inoperative

1. Check the fuse for the anti-theft system. Is it good?

____

____

2. Check the bulb of the security lamp. How did you check it and what did you find?

____

____

3. Check for voltage to the bulb. What did you find? If there was no voltage to the bulb, what is indicated?

____

4. Using the wiring diagram, locate the diodes in the system. Describe where they are.

____

____

5. With your DMM check each diode and summarize what you found.

____

____

6. Carefully check all of the ground circuits and connections in the system. Summarize what you found.

____

____

Problems Encountered

____

____

____

Instructor's Comments

____

____

____

— NOTES —

— NOTES —

— NOTES —

— NOTES —

— NOTES —

— NOTES —

— NOTES —

— NOTES —

— NOTES —

— NOTES —

— NOTES —

— NOTES —